Globalizing Responsibility

RGS-IBG Book Series

Published

Globalizing Responsibility: The Political Rationalities of Ethical Consumption
Clive Barnett, Paul Cloke, Nick Clarke and Alice Malpass

Domesticating Neo-Liberalism: Spaces of Economic Practice and Social Reproduction in Post-Socialist Cities
Alison Stenning, Adrian Smith, Alena Rochovská and Dariusz Świątek

Swept Up Lives? Re-envisioning the Homeless City
Paul Cloke, Jon May and Sarah Johnsen

Aerial Life: Spaces, Mobilities, Affects
Peter Adey

Millionaire Migrants: Trans-Pacific Life Lines
David Ley

State, Science and the Skies: Governmentalities of the British Atmosphere
Mark Whitehead

Complex Locations: Women's Geographical Work in the UK 1850–1970
Avril Maddrell

Value Chain Struggles: Institutions and Governance in the Plantation Districts of South India
Jeff Neilson and Bill Pritchard

Queer Visibilities: Space, Identity and Interaction in Cape Town
Andrew Tucker

Arsenic Pollution: A Global Synthesis
Peter Ravenscroft, Hugh Brammer and Keith Richards

Resistance, Space and Political Identities: The Making of Counter-Global Networks
David Featherstone

Mental Health and Social Space: Towards Inclusionary Geographies?
Hester Parr

Climate and Society in Colonial Mexico: A Study in Vulnerability
Georgina H. Endfield

Geochemical Sediments and Landscapes
Edited by David J. Nash and Sue J. McLaren

Driving Spaces: A Cultural-Historical Geography of England's M1 Motorway
Peter Merriman

Badlands of the Republic: Space, Politics and Urban Policy
Mustafa Dikeç

Geomorphology of Upland Peat: Erosion, Form and Landscape Change
Martin Evans and Jeff Warburton

Spaces of Colonialism: Delhi's Urban Governmentalities
Stephen Legg

People/States/Territories
Rhys Jones

Publics and the City
Kurt Iveson

After the Three Italies: Wealth, Inequality and Industrial Change
Mick Dunford and Lidia Greco

Putting Workfare in Place
Peter Sunley, Ron Martin and Corinne Nativel

Domicile and Diaspora
Alison Blunt

Geographies and Moralities
Edited by Roger Lee and David M. Smith

Military Geographies
Rachel Woodward

A New Deal for Transport?
Edited by Iain Docherty and Jon Shaw

Geographies of British Modernity
Edited by David Gilbert, David Matless and Brian Short

Lost Geographies of Power
John Allen

Globalizing South China
Carolyn L. Cartier

Geomorphological Processes and Landscape Change: Britain in the Last 1000 Years
Edited by David L. Higgitt and E. Mark Lee

Forthcoming

Spatial Politics: Essays for Doreen Massey
Edited by David Featherstone and Joe Painter

The Improvised State: Sovereignty, Performance and Agency in Dayton Bosnia
Alex Jeffrey

In the Nature of Landscape: Cultural Geography on the Norfolk Broads
David Matless

Learning the City: Knowledge and Translocal Assemblage and Urban Politics
Colin McFarlane

Fashioning Globalization: New Zealand Design, Working Women and the 'New Economy'
Maureen Molloy and Wendy Larner

Globalizing Responsibility

The Political Rationalities of Ethical Consumption

Clive Barnett, Paul Cloke,
Nick Clarke and Alice Malpass

⊛WILEY-BLACKWELL

A John Wiley & Sons, Ltd., Publication

This edition first published 2011
© 2011 Clive Barnett, Paul Cloke, Nick Clarke and Alice Malpass

Blackwell Publishing was acquired by John Wiley & Sons in February 2007. Blackwell's publishing program has been merged with Wiley's global Scientific, Technical, and Medical business to form Wiley-Blackwell.

Registered Office
John Wiley & Sons Ltd, The Atrium, Southern Gate, Chichester, West Sussex, PO19 8SQ, United Kingdom

Editorial Offices
350 Main Street, Malden, MA 02148-5020, USA
9600 Garsington Road, Oxford, OX4 2DQ, UK
The Atrium, Southern Gate, Chichester, West Sussex, PO19 8SQ, UK

For details of our global editorial offices, for customer services, and for information about how to apply for permission to reuse the copyright material in this book please see our website at www.wiley.com/wiley-blackwell.

The right of Clive Barnett, Paul Cloke, Nick Clarke and Alice Malpass to be identified as the authors of this work has been asserted in accordance with the UK Copyright, Designs and Patents Act 1988.

Library of Congress Cataloging-in-Publication Data

Globalizing responsibility : the political rationalities of ethical consumption / Clive Barnett ... [et al.].
      p.   cm. – (RGS-IBG book series)
   Includes bibliographical references and index.
   ISBN 978-1-4051-4558-9 (hardback)
1. Consumption (Economics)–Moral and ethical aspects.   2. Social justice.   I. Barnett, Clive.
   HC79.C6G583 2010
   174–dc22

                                                    2010038163
A catalogue record for this book is available from the British Library.

Set in 10/12pt Plantin by SPi Publisher Services, Pondicherry, India

01   2011

Contents

Series Editors' Preface   vii
Preface and Acknowledgements   viii

1   **Introduction: Politicizing Consumption
    in an Unequal World**   1
    1.1   The Moralization of Consumption   1
    1.2   Justice, Responsibility and the Politics of Consumption   4
    1.3   Relocating Agency in Ethical Consumption   11
    1.4   Problematizing Consumption   19

Part One   Theorizing Consumption Differently   25

2   **The Ethical Problematization of 'The Consumer'**   27
    2.1   Teleologies of Consumerism and Individualization   27
    2.2   Theorizing Consumers as Political Subjects   33
    2.3   The Responsibilization of the Consumer   39
    2.4   What Type of Subject Is 'The Consumer'?   44
    2.5   Does Governing Consumption Involve
          Governing the Consumer?   48
    2.6   The Ethical Problematization of the Consumer   52
    2.7   Conclusion   58

3   **Practising Consumption**   61
    3.1   The Antinomies of Consumer Choice   61
    3.2   Theorizing Consumption Practices   64
    3.3   Problematizing Choice   70
    3.4   Articulating Background   77
    3.5   Conclusion   81

4 Problematizing Consumption   **83**
   4.1 Consumer Choice and Citizenly Acts   83
   4.2 Articulating Consumption and the Consumer   85
   4.3 Mobilizing the Ethical Consumer   90
   4.4 Articulating the Ethical Consumer   97
   4.5 Conclusion   107

Part Two   Doing Consumption Differently   **111**

5 Grammars of Responsibility   **113**
   5.1 Justifying Practices   113
   5.2 Researching the (Ir)responsible Consumer   117
   5.3 Versions of Responsibility   124
   5.4 Dilemmas of Responsibility   137
   5.5 Conclusion   149

6 Local Networks of Global Feeling   **153**
   6.1 Locating the Fair Trade Consumer   153
   6.2 Re-evaluating Fair Trade Consumption   155
   6.3 Managing Fair Trade, Mobilizing Networks   163
   6.4 Doing Fair Trade: Buying, Giving, Campaigning   170
   6.5 Conclusion   179

7 Fairtrade Urbanism   **181**
   7.1 Rethinking the Spatialities of Fair Trade   181
   7.2 Re-imagining Bristol: From Slave Trade to Fair Trade 185
   7.3 Putting Fair Trade in Place   189
   7.4 Fair Trade and 'The Politics of Place Beyond Place' 191
   7.5 Conclusion   195

8 Conclusion: Doing Politics in an Ethical Register   **198**
   8.1 Beyond the Consumer   198
   8.2 Doing Responsibility   200

Notes   203
References   210
Index   227

Series Editors' Preface

The RGS-IBG Book Series only publishes work of the highest international standing. Its emphasis is on distinctive new developments in human and physical geography, although it is also open to contributions from cognate disciplines whose interests overlap with those of geographers. The Series places strong emphasis on theoretically informed and empirically strong texts. Reflecting the vibrant and diverse theoretical and empirical agendas that characterize the contemporary discipline, contributions are expected to inform, challenge and stimulate the reader. Overall, the RGS-IBG Book Series seeks to promote scholarly publications that leave an intellectual mark and change the way readers think about particular issues, methods or theories.

For details on how to submit a proposal please visit:
www.rgsbookseries.com

Kevin Ward
University of Manchester, UK

Joanna Bullard
Loughborough University, UK

RGS-IBG Book Series Editors

Preface and Acknowledgements

This book is one of the outcomes of a three-year research project entitled *Governing the Subjects and Spaces of Ethical Consumption*, funded by the Economic and Social Research Council (ESRC) and the Arts and Humanities Research Council (AHRC) (Grant Number RES 143250022).[1] The project was part of the ESRC/AHRC *Cultures of Consumption Programme*.[2] The argument developed in the book draws on empirical research conducted in and around the city of Bristol in the south-west of England, as well as research on national-level policy, NGOs and business organizations.

We would like to thank the School of Geographical Sciences at the University of Bristol for hosting the project throughout its duration, including Jonathan Tooby for his assistance in photographing ethical consumption activities around the city. We would like to thank all of the participants in the *Cultures of Consumption* Programme, from whom we have learnt an enormous amount, and in particular the Programme Director, Frank Trentmann. Thanks also to Jessica Pykett for her contribution to the research project. The project benefited enormously from the input of the members of an advisory panel, Tracey Bedford, Rob Harrison, Roger Levett, Terry Newholm and Allan Williams, and we thank each of them as well as all of the participants in two end-of-project events, held at the Open University Regional Centre in Bristol and at the Royal Geographical Society in London in 2005, for their always helpful and critical feedback on our unfolding research plans and analysis. We would like to thank the CREATE Centre in Bristol and the Fairtrade Foundation for their help in hosting the *Taste of Life* photography exhibition in June 2004, and Siobhan Wall for bringing her exhibition *The Clothes She Wears* to the RGS in 2005. Finally, we would like to thank all those people in and around Bristol and beyond who agreed to take part in this project as research subjects, whether as interviewees, focus group participants or informal contacts.

At Wiley-Blackwell, we would like to thank Jacqueline Scott, Liz Cremona and Tom Bates for their help, and patience, with the preparation of this book. Thanks also to Kevin Ward, as Series Editor of the RGS-IBG Book Series, for his encouragement and advice, and thanks too to two anonymous reviewers of the original manuscript for helping us clarify our argument.

Some of the material published here has been published previously in different contexts. Parts of Chapter 2 include material from Barnett, C., Cloke, P., Clarke, N. and Malpass, A (2008) 'The elusive subjects of neoliberalism: beyond the analytics of governmentality', in *Cultural Studies* 22, 624–653, reproduced with permission of Taylor and Francis Group. Parts of Chapter 4 include material from Clarke, N., Barnett, C., Cloke, P. and Malpass, A. (2007) 'Globalizing the consumer: doing politics in an ethical register', *Political Geography* 26, 231–249, reproduced with permission of Elsevier. Chapter 6 is a revised and extended version of an article that first appeared as Clarke, N., Barnett., C., Cloke, P. and Malpass, A. (2007) 'The political rationalities of fair-trade consumption in the United Kingdom', *Politics and Society* 35, 583–607, reproduced by permission of Sage Publications Ltd. Chapter 7 is a revised version of an article that first appeared as Malpass, A., Cloke, P., Barnett, C. and Clarke, N. (2007) 'Fairtrade urbanism: the politics of place beyond place in the Bristol Fairtrade City campaign', *International Journal of Urban and Regional Research*, 31, 633–645, reproduced by permission of Wiley-Blackwell.

Every effort has been made to trace copyright holders and to obtain their permission for the use of copyright material. The publisher apologizes for any errors or omissions in the above list and would be grateful if notified of any corrections that should be incorporated in future reprints or editions of this book.

Chapter One

Introduction: Politicizing Consumption in an Unequal World

1.1   The Moralization of Consumption

In contemporary debates about climate change, human rights, social justice, sustainability and public health, patterns of everyday consumption are commonly identified as both a source of harm and as a potential means of addressing various problems. In turn, consumers are routinely challenged to change their behaviour through the exercise of responsible choice. In this book, we develop a genealogical analysis of the institutional, organizational and social dynamics behind the growth in ethical consumption practices in the United Kingdom. We theorize this phenomenon in terms of the *problematization* of consumption and consumer choice. We argue that the emergence of ethical consumption is best understood as a political phenomenon rather than simply a market response to changes in consumer demand. By this, we mean that it reflects strategies and repertoires shared amongst a diverse range of governmental and non-governmental actors. The emergence and growth of contemporary ethical consumption is, we propose, indicative of distinctive forms of political mobilization and representation, and of new modes of civic involvement and citizenly participation. In developing this argument, we seek to counter the common view that the emergence of ethical consumption activities is a sign of the substitution of privatized acts of consumer choice for properly political forms of collective action. In order to move beyond the terms of this negative evaluation of 'consumerism', we argue that it is necessary to displace 'the consumer' from the centre of analytical, empirical and critical attention.

In this book we develop the argument that the emergence of ethical consumption should be understood as a means through which various actors

Globalizing Responsibility: The Political Rationalities of Ethical Consumption by Clive Barnett, Paul Cloke, Nick Clarke and Alice Malpass   © 2011 Clive Barnett, Paul Cloke, Nick Clarke and Alice Malpass

seek to 'do' politics in and through distinctively ethical registers. Above all, it is the register of responsibility that is prevalent in the diverse activities that make up the field of ethical consumption. We argue that ethical consumption campaigning is a form of political action which seeks to articulate the responsibilities of family life, local attachment and national citizenship with a range of 'global' concerns – where these global concerns include issues of trade justice, climate change, human rights and labour solidarity. In short, we are interested here in understanding how ethical consumption campaigning seeks to 'globalize responsibility'.

In developing our argument, we take our distance from the two dominant social science traditions of thought about the politics of consumption. In the first, consumption serves as a privileged entry point for thinking about the attenuated moral horizons of modern life. In this paradigm, Marx's account of commodity fetishism is reframed as a hypothesis about the deleterious effects of affluent consumers having no knowledge about the origins of the goods that they consume. On this view, responsible action requires the development of cognitive maps that connect spatially and temporally distanciated actions and their consequences through the provision of explanatory knowledge. The moral charge of research on commodity chains and value chains lies in the claim that by reconnecting locations of production, networks of distribution and acts of consumption, the alienating effects of modern capitalism can be exposed. Behind this style of analysis is the assumption that the secret to motivating practical action lies in helping people to recognize their entanglement in complex networks of commodification and accumulation.

In a second tradition of research on the politics of consumption, the emphasis is on asserting the skilled, active and creative role of consumers and consumption activities. Research in sociology, anthropology, cultural studies and human geography has demonstrated that everyday commodity consumption is a realm for the actualization of capacities for autonomous action, reflexive monitoring of conduct and the self-fashioning of relationships between selves and others. Here, consumption is reframed as a field in which ordinary people resist, subvert and creatively appropriate dominant cultural registers of consumerism.

From the perspective of the first of these traditions, the moral issue raised by commodity consumption is the imperative of attending to the consequences of extended networks of production and distribution that people are entangled in by virtue of their actions as consumers (see Wilk 2001). This view is often associated with the assumption that consuming more responsibly is equivalent to consuming less. From the perspective of the second tradition, however, the central moral issue is the acknowledgement of the ways in which consumption offers people opportunities to determine the types of selves and the types of relationships they wish to cultivate. This perspective is much more attuned to appreciating the importance of objects of consumption to practices of self-making (see Miller 2001c). And these two views are not necessarily opposed of course. The moral force of demonstrating the chains of consequence into which consumers' identities are woven tends to assume that, once informed about these consequences, people have the capacity to take responsibility for changing their consumption activity accordingly, in order to minimize environmental impacts, to boycott unethical companies, or to support fair trade or organic product ranges.

Both of these styles of critical social science stand in a longer tradition of *moralizing* about consumption. Hilton (2004) observes that from the eighteenth century through to the mid-twentieth century, consumption itself tended to be subjected to (largely negative) moral judgement. The rise of modern consumer politics in Europe and North America after 1945 represented a *demoralization* of consumption, in so far as this politics focused on the benefits and risks associated with specific products in a context in which generalized mass consumption was considered a norm. In this respect, Hilton (ibid.: 119) suggests that the late twentieth century and early twenty-first century saw 'a discernible trend to remoralize the market through issues of ethical consumerism and globalization'. In this context '[m]oralities of consumption might therefore be re-emerging as globalized critiques of the discrepancies in northern affluence and southern poverty' (ibid.: 120).

The dominant motif of the contemporary remoralization of consumption is the revival of a long-standing tradition of opposing the egoistic, hedonistic, self-interested imperatives of the consumer to the civic virtues of the citizen. Schudson (2007: 237) has observed that 'a lot of criticism of consumer culture has been moralistic, judgemental, intolerant, condescending, and, perhaps, muddled'. The muddle involved in the criticism of consumer culture is most evident, he suggests, in this opposition of consumer to citizen. It is a trope that 'offers a narrow and misleading view of consumer behaviour as well as an absurdly romanticized view of civic behaviour' (ibid.: 238). The idea that commodity consumption and consumerism are irredeemably individualistic, irresponsible and apolitical is in need of revision, not least in light of the centrality of consumer activism to histories of modern citizenship, civil society and welfare (e.g., Hilton 2003, 2008; Trentmann 2008). As Schudson (2006: 203) suggests, 'in an age of environmentalism, consumer boycotts, and political regulation of the safety of cars and toys and pajamas', the assumption that the world of consumption and the world of politics are guided by diametrically opposed values is 'ripe for reconsideration'. With this in mind, understanding the globalization of responsibility through discourses of 'ethical' consumerism requires us to adopt what Schudson calls a *post-moralistic* approach to understanding the contemporary politicization of consumption. But this requires us to take a detour through some recent moral and political philosophy to better grasp what is at stake in thinking seriously about the concept of responsibility and its relationship to the contemporary politics of consumption.

1.2   Justice, Responsibility and the Politics of Consumption

If a great deal of academic analysis of consumption is implicitly if not explic-itly moralistic, then it is also the case that much of this analysis tends to presume that the moral values associated either negatively or positively with consumption are self-evident. Moralizing about consumption depends on simplifying a complex range of practices, processes and relations. Seen from one angle, the active, assertive consumer of cultural studies lore is able to maintain multiple personal relations of care and love through the purchase, exchange and use of commodities. Seen from another angle, they are complicit in the reproduction of systematic inequalities of global wealth, environ-mental damage and human rights abuses. And it is the latter perspective that has attracted most sustained attention amongst scholars interested in connecting consumption to the concerns of moral and political philosophy (see Crocker and Linden 1998).

Our starting point is that reasoning about issues of responsibility and con-sumption should not be reduced to a causal calculation of causes and effects. Nor should we necessarily frame these issues in the purely 'ethical' terms of worrying about how affluent consumers in the West should best discharge their obligations to assist those less fortunate than themselves. These two frames – in which responsibility is reduced to a matter of causality and/or a matter of assisting those less fortunate – are the primary registers in which issues of responsibility have been discussed in human geography's so-called 'moral turn' (see Smith 2000 for a review of this field). But we need to keep in view the close proximity between issues of responsibility and questions of justice. The political dimension of justice is hardly absent from this set of debates in human geography, but there is a persistent tendency to think of values such as justice, care or responsibility as externally generated criteria against which the world should be judged and by which action should be guided. In contrast, our concern here is with developing an account of responsibility and justice understood as normative modalities through which practices unfold in the world (see Boltanski and Thévenot 2006).

Debates in political philosophy about *global distributive justice* provide an entry point for framing the relationship between responsibility and consumption. Thomas Pogge (1994, 2001) has argued that rather than rea-soning about obligations to those less fortunate than oneself from the perspective of a potential helper, it is more appropriate to acknowledge that affluent citizens of the West stand in the position of supporters and benefi-ciaries of global institutional systems that contribute to the impoverishment and disenfranchisement of distant others. Pogge's point is that questions of global responsibility are not merely matters of personal morality; they are also issues of *justice*. Or, from a related but distinct position, Onora O'Neill (2000) argues that equal moral status should be afforded to 'distant others' because, in everyday activities, their status as moral agents is taken for granted. Like Pogge, O'Neill is making an argument not just about moral responsibility, but about equality and justice. Both of these positions are part of a broader field of debates in which the principle of egalitarianism as the core value of justice is framed in terms of particular understandings of responsibility (see Hurley 2003).

The arguments of Pogge and O'Neill are part of a broader philosophical debate about the degree to which the egalitarian theory of justice developed and defended by John Rawls (1972) can be applied to transnational processes and the global scale (see Tasioulas 2005). The pivotal issue in these debates concerns the question of just what range of activities should be evaluated by an egalitarian theory of justice. Rawls (1972: 7–11) originally argued that 'the subject of justice' should first and foremost be thought of as the institutions of society which sustain deep and pervasive inequalities – what he called 'the basic structure of society'. The basic structure included 'the political constitution and the principal economic and social arrangements' (ibid.: 7). On this view, then, the primary subject of justice is 'the way in which the major social institutions distribute fundamental rights and duties and determine the division of advantages from social cooperation' (ibid.). In making this argument for 'the basic structure of society' as the primary subject of evaluations of justice, Rawls was imposing a restriction on the range of activities that theorizing about 'social justice' should be expected to address:

> Many different kinds of things are said to be just and unjust: not only laws, institutions, and social systems, but also particular actions of many kinds, including decisions, judgments, and imputations. We also call the attitudes and dispositions of persons, and persons themselves, just and unjust. (Ibid.)

Rawls did not consider this broader range of activities to be a primary concern of a theory of social justice. For him, these were matters of ethical judgement rather than evaluations of justice (cf. O'Neill 1996).

Philosophical debates about global distributive justice focus on the question of whether it is plausible to think in terms of a 'global basic structure', and therefore whether it is appropriate to extend and revise principles of egalitarian justice to this scale of evaluation (Pogge 1994). Rawls (1999) himself thought that principles of distributive justice could not be extended in this way, and affirmed instead a somewhat paternalist 'duty of assistance' to other peoples. But cutting across positions on this issue of the *scope of justice* there is a debate concerning whether the restriction of questions of justice to the basic structure is actually justifiable. In a radical endorsement of the principle that 'the personal is political', G. A. Cohen (2000, 2008) argues against Rawls that this restriction should be lifted, so that 'non-coercive' structures such as conventions, social ethos and personal choices also fall under the evaluation of egalitarian principles of justice. For Cohen, principles of justice necessarily make a claim on personal conduct; they require the *site* of justice to be extended beyond 'political' fields all the way into matters of 'ethics'. Against this view, while arguing that the scope of distributive justice principles should be extended globally, Pogge (2000) argues against the idea that egalitarian criteria for assessing institutional structures should also apply to a morality that governs personal conduct as well. In his view, this 'monist' position will actually hinder the development of an overlapping consensus on issues of global distributive justice.

These debates in political philosophy about the *scope* (global or domestic) and *site* ('coercive' institutions of the basic structure and/or 'non-coercive' fields of personal conduct and social ethos) of egalitarian justice are played out in practice in the contemporary politicization of consumption. As we have already suggested, commodity consumption is routinely presented as implicated in wider networks of global inequality and environmental harm. A recurring theme of public debates about what to do about consumption is the problem of where effective agency for changing consumption lies. Is it the responsibility of 'everyone', as consumers, to 'do their bit' and 'play their part' in reducing the unjust, destructive, unsustainable consequences of consumption? Or does attention need to be focused on structural factors, such as the regulation of markets, the monitoring of production and distribution systems, or re-gearing international financial and trade regimes? Arguing the former position might well lead to charges that this lets the real culprits off the hook, as well as imposing unreasonable burdens on socially differentiated groups of consumers. Arguing the latter position might elicit the charge that this passes the buck – that, as citizens, people should be more responsible about exercising consumer choice. In Chapter 5 of this book, we show how both of these sentiments find expression in the ordinary forms of reasoning that are bought to bear by 'consumers' about the responsibilities they are supposed to discharge through their everyday consumption choices. Philosophical tensions concerning the scope and site of justice are, then, very real matters of public debate, campaign strategy and policy design in the contemporary politics of consumption.

The contribution to this set of philosophical debates that most informs our analysis here is that of the feminist political philosopher Iris Marion Young. Young uses the contemporary politicization of consumption as the real-world example with which to work through these questions of the relationship between justice and responsibility. In her work on political responsibility, Young deploys the example of anti-sweatshop campaigns, particularly the movement that developed in the United States from the mid-1990s around this issue, to clarify the types of critical reasoning that might be applied to issues of global injustice. Young's (1990) own retheorization of justice stands in critical relation to the Rawlsian heritage, not least by extending the definition of the basic structure to include a wide range of non-distributive issues, such as the social division of labour, structures of decision-making power and processes of cultural normalization' (see Young 2006). This is a *deepening* of the definition of the basic structure, as distinct from the *extension* of the scope of the Rawlsian principle of the basic structure that Pogge recommends or the *generalization* that Cohen recommends. Against the type of position developed by Cohen, Young reaffirms the Rawlsian principle of two levels of moral evaluation: 'one to do with individual interaction and the other to do with the background conditions within which that action takes place' (2006: 91). She affirms that questions of justice refer primarily to the latter level:

Theorizing justice should focus primarily on the basic structure, because the degree of justice or injustice of the basic structure conditions the way we should evaluate individual interactions or rules and distributions within particular institutions. (Ibid.)

Young therefore refuses to collapse institutional analysis into the analysis of individual interactions. Young's commitment to the idea that the primary subject of justice is the basic structure, suitably extended *and* deepened, helps us to appreciate the task she pursues in her theorization of the modalities of political responsibility disclosed by the anti-sweatshop movement. Her account of political responsibility is a response to the challenge of developing a point of view that can encompass 'the accumulated consequences of the actions of millions of mediated individuals' (ibid.: 96). Young's own response focuses attention on how this challenge is being practically met through innovative forms of transnational mobilization and campaigning.

Political responsibility emerges in Young's analysis as a theme in which questions of justice are articulated with the evaluation of individual-level conduct and interaction in a non-reductive way. Young (2004) develops what she calls a social connection model of political responsibility, in which responsibility is understood to arise from the ways in which different actors are implicated in structural social processes. Her starting point is a concern not to reproduce a discourse of blame and guilt by applying a single standard of justice to both social structures and individual action (Young 2003). Consumer campaigns often invoke the theme of collective responsibility in the effort to motivate individual behaviour change, implicitly falling back on a model in which responsibility is about being held liable for the consequences of one's actions. But staking claims about responsibility on a liability model of responsibility is, Young points out, likely to be counterproductive, and also risks reproducing injustices of its own. In seeking to motivate action, it is not enough to show that someone is connected through their everyday actions as a consumer to wider systems that reproduce harm. Quite the reverse in fact, since on their own, any one individual might not actually be able to do much about sweatshops on the other side of the world. In many arguments of this sort, an individual consumers' connection to labour exploitation or environmental destruction in distant places seems to depend on a whole chain of mediating causal linkages. These might persuade an individual consumer that their actions contribute, in small ways, to the reproduction of those harms. But they are just as likely to convince the consumer that on their own, as a consumer, there is not much they can do about it (see Allen 2008).

In contrast to this blame-focused understanding of responsibility, Young sets out an understanding of political responsibility that can negotiate between an undifferentiated claim of individual responsibility and an undiscriminating claim about collective responsibility. Young (2007: 179) calls this alternative a model of *shared responsibility*, one in which responsibility is distributed across complex networks of causality and agency (see also Kuper 2005; Barnett, Robinson and Rose 2008). The advantage of the concept of shared responsibility is that it allows a more discriminating analysis of the partial ways in which actors might understand themselves to be responsible, where this in turn is not just a matter of liability or blame but is closely related to an analysis of the capacity to act. Young argues (2007: 181–186) that there are varied 'parameters of reasoning' about responsibility that can be practically applied to link up questions of global justice and personal-political responsibility: responsibility can be understood along vectors of power, privilege, interest and collective ability.

Young's approach to political responsibility enables us to see that responsibility does not, in theory or practice, arise simply from being connected to events, people, places and processes. It is differentiated according to capacities that actors can bring to bear to change things. For example, the question of *power* is one crucial dimension of this sense of shared responsibility – it is important to be able to discern the degree and type of influence that different actors have to change a situation. But responsibility might also be differentiated by the issue of relative *privilege*. So, for example, even actors who cannot reasonably be thought of as causally responsible for sweatshop labour might be plausibly understood to benefit from or derive privileges from these conditions and practices. In short, it might be the case that those who gain privileges from patterns of harm done to others are also those who have most power to act to change these. But it is not necessarily the case. Benefiting from these patterns, causing them, and having the capacity to act to change them do not map easily onto a single location in social space – and an analysis of the politics of responsibility therefore needs to be attentive to how these different dimensions of responsibility are articulated together.

The main lesson we take from Young's work on political responsibility is that theorizing about *global* political responsibility requires more than just telling stories about spatially extensive networks of connection and entanglement.

It also requires avoiding simple assertions of collective responsibility over individual responsibility. Thinking in terms of shared responsibility, as distinct from collective responsibility, leads us to think about the ways in which the power to influence events is widely distributed. In short, responsibility has different forms and is shared among different actors. It is also likely to be motivated by concerns for different moral 'goods'. So, for example, Micheletti and Stolle (2007) re-describe the anti-sweatshop movement from which Young's analysis draws in terms of the articulation of different movement actors who seek to mobilize individuals along four pathways. In this case, the actors include trade unions, anti-sweatshop associations, international humanitarian organizations, and Internet activists; and the individuals are mobilized as supporters of causes, as a critical mass of shoppers, as agents of corporate change and as ontological agents of societal change.

Young's account of the modalities of responsibility disclosed by campaigns for global labour solidarity does more than tell a simple geographical story about the responsibilities that people have by virtue of being connected into wider spatial systems. Her account of political responsibility stresses questions of power and privilege as well as simply connection: some actors bear more responsibility by virtue of having greater capacity to act; some by virtue of being relatively privileged by their position in unequal systems of social relations. In short, taking responsibility is not just an individuated action taken by a single person or by some collective agent. It is theorized in terms of how distributed actions join actors together, feeding into wider networks of cooperation that reach out and influence events elsewhere.

Young's philosophical reconstruction of the modalities of political responsibility assists us in the task of developing a 'post-moralist' approach to understanding the contemporary politicization of consumption through ethical registers. In this book, we seek to develop a complementary analysis, drawing inspiration from Young's framework in order to understand the practical ways in which the politicization of consumption in Western capitalist democracies can be understood as responding to broadly shared concerns about the possibility of living responsibly in a highly unequal world (see also Massey 2006; Jackson et al. 2009).

In emphasizing how responsibility is being discursively and practically globalized through ethical consumption campaigning, we want to avoid falling into the trap of presuming that consumption has become an ethically charged, politically contentious arena simply in response to secular transformations in the nature of 'modernity'. Our emphasis in this book is on understanding the ways in which the widespread turn to consumer-based forms of mobilization among campaign groups, NGOs, social movements, as well as by policy makers, is the result of the strategic search for effective agents of change in an increasingly complex 'globalized' world economy (Littler 2005). In short, we are interested in understanding how different agents make 'global responsibility' into both a problem and a possibility.

The politicization of consumption in the register of responsibility is not as a substitute for other forms of political action. Rather, this politicization seeks to link practices and social relations of consumption to the transformation of broader systems and social relations of production, distribution and trade (Murray 2004). Kate Soper (2006) argues that if universal criteria of basic needs are to be met on a global scale, then the required changes to patterns of material affluence in Western societies are not likely to be generated through moralistic demands to consume less or consume more responsibly. What is required is the development of alternative political imaginaries which help in redefining understandings of needs, pleasure and enjoyment. The range of practices we examine in this book under the broad topic of 'ethical consumption' deserve, we think, to be considered as practical experiments in developing just these sorts of alternative imaginaries. And they are also experiments in developing the sort of distributed practices of shared responsibility outlined by Young.

We follow Leyshon, Lee and Williams in suggesting that fair trade consumption, organic food networks, sustainable consumption initiatives and other examples of alternative economic activity should be seen as 'practical, day-to-day experiments in performing the economy otherwise' (Leyshon et al. 2003: 11). But it might be wise to avoid the idea that experiments of this sort must somehow escape entanglement in a supposedly all-encompassing capitalist monolith to deserve attention as repositories of critical alternatives (see Luetchford and De Neve 2008). To adopt this 'critical' axiom is lose sight of the political significance of day-to-day performances of values and commitments by deploying a totalizing vision of higher level structural transformation. In insisting on analysing these alternative economic practices as political we are keen to avoid premature celebration just as are we are keen to resist the temptations of easy 'critique'. We are concerned in this book with underscoring the extent to which the alternative economic practices involved in the growth of ethical consumption are embedded in organized, strategic campaigns which are focused as much on mobilizing support and making claims of representation as they are on simply getting people to buy this or that brand of coffee or boycott this or that make of training shoe. In short, we start off by recommending Rob Harrison's characterization of what he calls 'ethical consumerism' as a movement. Harrison is one of the founding members of the Ethical Consumer Research Association and its magazine Ethical Consumer (discussed further in later chapters), and a leading figure in this movement in the UK and internationally. For Harrison, this movement grows out of the widespread adoption of 'market campaigns' by pressure groups and campaign organizations. In market campaigns, 'persuading consumers to act ethically is often just one element of a broader campaign which may involve other activities such as shareholder actions, political lobbying, pickets and non-violent direct action' (Harrison 2005: 55).

1.3   Relocating Agency in Ethical Consumption

Academic and activist discourses of capitalist globalization and rampant neoliberalism have provoked interest in the economic and political potential of various 'alternative economic spaces' (Leyshon *et al.* 2003). This includes research on the growth of ethical finance, alternative food networks, the social economy, and alternative trading systems (e.g., Carter and Huby 2005; Amin *et al.* 2002; Hughes 2005; Whatmore and Clark 2008). Ethical consumption is part of this broader flourishing of economic experimentation. In the United Kingdom, the market for 'ethical' consumer products has grown steadily since the early 1990s. Consumer expenditure on 'ethical' products in the UK almost tripled between 1999 and 2009, growing from £13.5 billion to £36 billion. This includes everything from fair trade and organic food to eco-friendly travel, energy efficient boilers to rechargeable batteries – and therefore reflects a range of 'global' issues, from trade justice to climate change.[1] While this growth coincides with a decade of escalating consumer spending, fuelled by credit and rising property prices, organizations involved in promoting this market suggest that the increase is also likely to prove resilient despite the onset of economic recession.[2]

The growth of academic research on ethical consumption has matched its economic growth (for a thorough review, see Newholm and Shaw 2007). A feature of both academic and popular discussions of the growth of ethical consumption is the widespread assumption that 'the consumer' is the key agent of this process. There is a burgeoning literature in economics and management studies on business ethics and corporate social responsibility. This work understands ethical consumption primarily in terms of the role that information plays as the medium through which the ethical preferences of consumers and the ethical records of businesses are signalled in the marketplace (e.g., Bateman; Fraedrich and Iyer 2002). From this perspective, the development of appropriate informational strategies (marketing, advertising, labelling and branding) will assist in overcoming market failure.

Academic research on ethical consumption as a form of political action is also often underwritten by the dual assumptions that providing information to consumers about the conditions of production and distribution of commodities is central to changing consumer behaviour, and that knowledge is the key to putting pressure on corporations and governments. The literature on consumer-oriented activism and policy – such as fair trade campaigns, sustainable consumption and ethical trade audits – often presumes that publicity is the primary means of acting on the conduct of individualized consumers and corporate actors alike. The literature which argues that ethical consumption and political consumerism are distinctively new forms of political practice tends to reproduce the idea that shopping is a key vector of action in a 'post-political age', narrowing the focus of attention on 'the consumer' (e.g., Cook *et al.* 2006; Hooghe and Micheletti 2005). This sort of framing of ethical consumption reproduces generalizing narratives in which 'traditional' forms of participation – party membership, voting – are supposed to be in terminal decline, and are being replaced by more individualized forms of action, for which buying or boycotting as a 'consumer' has become the paradigm (e.g., Stoker 2006).

This assumption is still at work even in academic literature that sets out to explain the splicing together of consumption practices and various campaign issues in non-reductive ways, and to justify this phenomenon as one worthy of serious academic study. The most important strand of research in this regard is literature that develops the concept of *political consumerism*. We discuss this literature in more detail in Chapter 2. This concept reproduces the assumption that 'the consumer' is a key agent of social change and that shopping is a medium of political action (e.g., Micheletti and Follesdal 2007). The same assumption frames critical analysis of strategies of 'shopping for change', the limitations of which are found to lie in the individualized, consumerized forms of activity through which this sort of action unfolds (e.g., Bryant and Goodman 2004; Guthman 2007; Low and Davenport 2007; Seyfang 2005; Varul 2008; Watson 2007). There remains a deep and intense suspicion that consumerized forms of public mobilization threaten to undermine or substitute for authentic, properly political collective activity.

The academic framing of ethical consumption as irredeemably tainted by its association with the cultural registers of consumerism is reflected in public discourse about the phenomenon, in which the shared assumption is that consumer motivations are the primary driver of this growth. So, for example, *The Economist* magazine has acknowledged the growing importance of the market in green, organic, and fair trade goods and services.[3] While paying lip service to the 'noble aims' of this market sector, it expresses suspicion of the idea that markets and consumer choice can or should serve as effective mediums of 'ethical' or 'political' goals – it recommends voting as the preferred means for this.[4] Likewise, left-liberal criticism of the growth of markets in broadly defined 'ethical' goods and services routinely alights upon the limitations of consumerism as a means of bringing about meaningful political change; expressing doubts that shopping can save the planet routinely slides into the lament that consumerism is replacing citizenship as the primary form of public engagement.[5]

In this book, we contribute to a reconceptualization of ethical consumption which challenges the assumption that 'the consumer' is the primary agent of change in efforts to politicize consumption practices (see Clarke 2008). In one sense, the focus on consumers in research on fair trade, alternative food networks, trade justice and environmental politics has opened out new perspectives on the personal ethical sensibilities that are addressed in these fields. However, this fixation on the agency of 'the consumer' – either positively or negatively evaluated – often fails to credit the role of campaign organizations as prime movers in the politicization of consumption. We suggest that in order to understand both the range of roles and motivations people bring to their engagements with ethical consumption campaigns, or the ways in which campaigns seek to enrol supporters, the concept of 'the consumer' might not throw much explanatory light on the set of processes involved in the growth of the variety of alternative economic practices subsumed under the name 'ethical consumption'.

So, just what is *ethical consumption*? What is known as ethical consumption in the UK bears a close resemblance to what European scholars and activists have called 'political consumerism' (Micheletti *et al.* 2004; Stolle *et al.* 2005), and part of the phenomenon that Littler (2008) has called 'radical consumption'. Ethical consumption is in important respects distinctive from anti-consumerist movements (Littler 2005; Zavestoski 2002) such as the voluntary simplicity movement (Cherrier and Murray 2002; Shaw and Newholm 2002) or 'No Logo' forms of anti-globalization campaign (Klein 2000). Rather than rejecting the persona of 'consumer', ethical consumption campaigning represents a distinctive strategy for connecting the politics of consumption with the practices of being a discerning, choosey, responsible consumer. It is therefore more aligned with slow food movements (Andrews 2005), although often more populist in its methods and objectives, and more closely aligned to development and green political movements. Ethical consumption campaigning is also distinct from the related and growing area of ethical investment (Carter and Huby 2005). Ethical consumption campaigning seeks to embed altruistic, humanitarian, solidaristic and environmental commitments into the rhythms and routines of everyday life – from drinking coffee, to buying clothes, to making the kids' packed lunch. But it must also, we suggest, be analysed not simply in terms of the changes to patterns of consumption that it succeeds in generating. Ethical consumption, understood as an organized field of strategic interventions, seeks to use everyday consumption as a surface of mobilization for wider, explicitly political aims and agendas. Thus, it marks an innovation in modes of 'being political' (Isin 2002), one in which people are encouraged to recognize themselves as bearing certain types of *global* obligation by virtue of their privileged position as consumers, obligations which in turn they are encouraged to discharge in part by acting as consumers in 'responsible' ways. The sense of 'global' here is itself open to multiple interpretations in different campaigns – it encompasses not only activities premised on the assumption that consuming certain goods can assist distant actors or help in reshaping international trade, but also activities that seek to reshape highly localized practices in order to minimize 'impacts' or 'footprints' that contribute to broader environmental processes.

It is worth underscoring just how diverse the activities that fall within the field of 'ethical consumption' actually are. Micheletti (2003) suggests that what she calls 'political consumerism' involves some combination of three forms of action: boycotting; positive buying, or 'buycotting'; and 'discursive' action of various sorts, from culture jamming to publicizing working conditions in distant sweatshops, in which information about consumption is circulated. Even this simple categorization indicates some of the diversity and complexity of ethical consumption. The Ethical Consumer Research Association (ECRA), the leading ethical consumption campaign organization in the UK, works with a similar sort of categorization (Harrison 2008). For them, ethical consumption includes different forms of action: boycotting, positive buying, anti-consumer activity, buying the most ethical product in a particular sector, or relationship purchasing. In turn, different sorts of commodities might be the focus of ethical consumption activity (see Crane 2001): boycotted goods might include environmentally destructive products (e.g., aerosols) or high-profile corporations (e.g., Nestlé, Esso); positive buying, perhaps exemplified by fair trade, might focus on coffee or chocolate; best-in-sector buying depends heavily on labelling schemes, and might extend from food products to energy efficient washing machines; relationship buying might include vegetable box schemes or shopping on the local high street. And in turn, different sorts of economic actors help to facilitate ethical consumption: specialist corporations branded as 'ethical', such as The Body Shop; mainstream retailers, such as Gap, who in 2006 launched, alongside American Express, the *Product (RED)* range, in which global brands were used to raise awareness and money to help finance AIDS programmes in Africa; as well as a range of alternative business models that have emerged in the past two decades. And ethical consumption might be considered as distinct from, but related to, the growth of ethical banking and investment, and distinct from, but again might be articulated with, the development of corporate social responsibility initiatives.

It is also worth underscoring the point that the politics of ethical consumption activities is far from straightforward. It includes anti-consumerist and culture jamming practices that use media campaigning against 'consumerism' (see Littler 2005). It might draw on traditions of downshifting and voluntary simplicity (Shaw and Newholm 2002; Cherrier and Murray 2002). In these cases, it is consumerism and consumption *per se* that are targeted as objects of political action. The use of purchasing to positively support particular causes is distinct from this model of ethical consumption. In this case, all sorts of political campaigns now use consumer goods to raise funds and awareness. Some of these campaigns are explicitly focused on transforming economic practices of which consumption is part. The fair trade movement draws on the long-standing traditions of the cooperative movement. This movement is distinct from the union-based movements that have emerged since the 1990s, often focused on anti-sweatshop campaigns, and frequently focused on garment and textile sectors (see Hale and Wills 2005). These two organizational fields are, however, increasingly drawn together, not least as fair trade campaigning moves beyond a focus on food to other commodity sectors, such as fair trade cotton (e.g., Egels-Zandén and Hyllman 2006). And at the same time it is possible to detect a 'consumerist turn' in the strategies of anti-sweatshop campaigners (Johns and Vural 2000; Prasad *et al.* 2004; Traub-Werner and Cravey 2002). This is just part of a broader shift to adopt consumer-oriented strategies by a broad range of campaign organizations. In increasing numbers of sectors, labelling and monitoring of product ranges is becoming an established voluntary practice (Guthman 2007; Van den Burg *et al.* 2003). Ethical consumption also extends to recycling and waste campaigns, promoted by national governments and administered by local authorities (Bulkeley and Gregson 2009).

In short, just what counts as ethical consumption is itself open to some debate. On the one hand, ethical consumption might be defined in relation to particular objects of ethical concern. In this sense, consumption research defines a variety of issues as 'ethical', including environmental sustainability, health and safety risks, animal welfare, fair trade, labour conditions and human rights. On the other hand, this focus on consumption as a means of acting in an ethical way toward particular matters of concern extends across various forms of practice, including shopping, investment decisions and personal banking and pensions. The diversity of objects and practices that might constitute ethical consumption is underscored by considering the diversity of organizational forms that might be defined in this category: ethical trading organizations, lobby groups, fair trade campaign organizations, cooperative movements, consumer boycott campaigns, 'No-Logo' anti-globalization campaigns, local food markets and charity shops.

Even this short overview indicates the high degree of overlap between organizations, the diversity of strategies adopted and issues addressed, and the variability of scales at which ethical consumption activities operate. It is this complexity that leads us to adopt a genealogical style of analysis, one which seeks to identify the emergence of a shared set of strategic problematizations which seek to mobilize a diverse range of motivations, incentives and desires in developing large-scale forms of collective action that are able to induce meaningful change in the patterns of conduct of powerful economic and bureaucratic systems.

In adopting this genealogical perspective, we are seeking to avoid the moralism that characterizes much of the critical social science literature on ethical consumption. As we have already noted, some commentators identify new forms of political agency in the growth of ethically motivated exercise of conscious consumer choice. But this can just as easily be taken to confirm a shift away from active citizenship prompted by a broader process of individualization, and as evidence of a process of disengagement from political processes which are still taken to be the norm – deliberative, representative forms of public action. In a great deal of critical analysis, the benchmark of properly 'ethical' consumption remains the virtuous figure of the anti-consumer – voluntary simplicity and more recently the growth of the slow food movement easily come to serve as reference points for practices which are less compromised with markets and the culture of consumerism. The moralization of consumption therefore persists in analysis of the growth of ethical consumption, and is evident in the degree of suspicion directed towards consumer-based forms of social activity, often interpreted as indices of consumerist individualism or neoliberal hegemony. This suspicion is an index of the strong hold that a particular image of consumption continues to have on the academic imagination: consumption is often thought of as synonymous with conspicuous display and spectacle, if not outright hedonism; and it is routinely reduced to the discrete act of purchasing. This view of consumption is rooted in a theoretical and empirical imagination that runs from Veblen through Horkheimer and Adorno, to Galbraith, on to Baudrillard and through to Bauman. What gets lost from view in this picture is the ordinariness of much of the consumption that people do, and the politics of this ordinariness (Hilton 2007b). Our argument in this book is based on the assumption that it not possible to understand the dynamics of ethical consumption initiatives, whether from the strategic perspectives of campaign organizations or from the perspective of the people they seek to enrol into their projects, without appreciating the mundane and ordinary dimensions of consumption.

Our argument is that appreciating what Hilton (ibid.) calls the 'banality of consumption' goes hand in hand with bringing into view the ways in which ethical consumption is related to the changing dynamics of civic participation. In their review of changing patterns of civic engagement in the UK between 1984 and 2000, Pattie et al. (2003b: 631) found that 'people's participation in conventional political activities (such as voting, contacting a politician and attending a political meeting) has declined, whereas participation in consumption and contact politics (boycotting goods and contacting the media) has grown significantly'. The question that arises from this finding is whether this reflects a substitution of publicly oriented collective participation by identity-based, individually motivated and privatized forms of concern. The answer to this question depends in part on just how ethical consumption is understood and explained. As we have already suggested, in the burgeoning literature in economics and management studies on business ethics and corporate social responsibility, ethical consumption is understood primarily in terms of the effective consumer demand as the medium through which the ethical preferences of consumers and the ethical records of businesses are signalled in the marketplace. From this perspective, markets are perfectly capable of expressing people's ethical, moral or political preferences, just as long as appropriate informational strategies are developed (e.g., marketing, advertising, labelling and branding). This is also a background assumption in many policy initiatives on sustainability, in campaigning around the environment, and across the range of 'ethical' trading initiatives, where it often seems to be supposed that the main challenge is to provide people with more information in order to raise awareness of the consequences of their everyday consumption choices – then they will magically change their behaviour.

The information-led understanding of ethical consumption misses out a great deal of what actually shapes people's consumption activities (Hobson 2003; Jackson 2004). And it manages to reproduce a narrowly utilitarian conceptualization of ethical decision-making by consumers, companies and public organizations alike (see Barnett, Cafaro and Newholm 2005). We question whether this prevalent understanding of ethical consumption provides an accurate picture of people's 'ethical' motivations for engaging in such activities, and also question whether the 'political' significance of ethical consumption lies simply in the signalling of aggregated demands in markets for consumer goods. On both of these grounds, the information-based model of ethical consumption as a consumer-driven process interprets this phenomenon far too narrowly.

The main example of ethical consumption we deal with in this book is fair trade consumption. The public presence of the movement advocating fair trade has grown considerably in the UK since the early 1990s (see Anderson 2009). This presence is evident not only in the growth of the market for fair trade products, but also in terms of awareness of the issues at the heart of the fair trade movement. The relationship between growing the market for fair trade goods and public communication is a pivotal aspect of 'market campaigns' such as fair trade. In this respect, the 'mainstreaming' of fair trade, as products such as fair trade tea, coffee and chocolate or fair trade cotton clothing become established in leading high street retail chains such as Sainsbury's, the Co-op and Marks & Spencer, is a process whereby the public reach of fair trade messaging is expanded. For example, in 2009, leading chocolate brands including Cadbury's *Dairy Milk* and Nestlé's *Kit-Kat* also became accredited fair trade products. This represents one of the most significant projections of fair trade into the public realm. *Dairy Milk* is the UK's best-selling chocolate bar, meaning that the distinctive fair trade logo 'will be printed on the 300 million chocolate bars sold a year, and appear in 30,000 shops where the product is sold.'[6]

Fair trade is an international movement for social and environmental justice that develops alternative economic spaces of production, trade, retailing and consumption (Lamb 2008). The goals of the fair trade movement include improving the livelihoods and well-being of small producers; promoting development opportunities for disadvantaged groups of producers, in particular women and indigenous peoples; raising awareness among consumers of the negative effects of patterns of international trade on producers in the global South; campaigning for changes in the regulatory regimes governing international trade; and the protection and promotion of human rights. The international fair trade movement consists of certification agencies, producer organizations and cooperatives, trading networks and retailers. In this book, we use different aspects of fair trade consumption activity in the UK to elaborate on how ethical consumption campaigning scrambles some settled understandings of the 'who', 'where' and 'how far' of citizenly acts (Barnett, 2010).

Fair trade is not an example of anti-consumerism. It is the exemplary market campaign. Fair trade campaigners do not want people to stop drinking coffee, or to eat less chocolate. They want us to buy coffee or chocolate that is produced and distributed by organizations and networks that ensure the producers receive a fair price to give them a sustainable livelihood. Looked at in purely economic terms, however, the impact of fair trade is still only a pinprick on unequal patterns of world trade. But maybe that is missing the point. As we show in Chapter 6, for people actively involved in fair trade consumption activities, the point to 'doing' fair trade is not wholly about the aggregate market effect of lots of individualized purchases. Although the potential impact on producer communities is a significant factor in people's motivations, engagement with fair trade consumption is often a way of aligning quite abstract commitments with the routines and rhythms of everyday life. In turn, for the organizations behind the growth of fair trade, consumer-based activism is an important way of raising awareness about issues, establishing the legitimacy of their claims and the validity of their own arguments, and generating 'demonstration effects' in support of alternative trading models. In the UK, companies and organizations such as Traidcraft, the Fairtrade Foundation, Oxfam, Christian Aid or the Co-op are all trying to exert influence over governments and corporations. It is very important for them to be able to show that they have broad-based popular support for the sorts of changes that they are promoting. Being able to demonstrate a growth in sales of fairly traded products is, then, one way for these organizations to legitimize their standing in a wider public realm, as well as validating themselves to members and supporters.

In this book we use empirical analysis of fair trade campaigning and other ethical consumption activities to develop an understanding of the ways in which contemporary forms of consumer-oriented activism seek to provide pathways to participation for a wide range of social actors, whether as individuals differently placed in socio-economic or institutional relationships, or public and private organizations embedded in different forms of economic, educational or civic activity. Some of the most successful contemporary campaigns for social justice – around labour, human rights and environment – use consumer-oriented strategies to raise awareness, mobilize support, and exert pressure on powerful actors. The significance of ethical consumption campaigning needs to be assessed not primarily in terms of the aggregate impact of individualized consumer choices on overall market trends, but rather in terms of the reorientation of the ways in which the mobilization of support, resources and legitimacy for political campaigns is sustained.

From the perspective which we develop in this book, the emergence of ethical consumption is not best explained by people becoming more ethically aware by virtue of learning about the extended consequences of their actions, and nor are changes in behaviour just responses to being provided with information about alternatives. Rather, this trend is the outcome of organized efforts by a variety of collective actors to practically re-articulate the ordinary ethical dispositions of everyday consumption. There are two dimensions to this process of practical re-articulation. First, there is a process of *discursive* engagement with the frames of reference that already shape people's consumer behaviour. Campaigning materials and modes of address that are sensitive to the experiential horizons of ordinary consumers are more likely to succeed than those that suppose that consumers normally lead constricted, self-interested, moral lives. Second, there is the process of using various *devices* to actually enable people to readjust their consumption behaviour. Recycling boxes are an example of this; so is the labelling of food and other products; vegetable box delivery schemes are another; direct-debit donations to charities another. In analyzing both aspects of ethical consumption campaigning in this book, we will see that the 'the consumer' is not so much a locus of sovereignty and agency, as it is a rhetorical figure and point of identification only contingently related to the politicization of consumption.

1.4 Problematizing Consumption

We have already indicated that there is a range of academic approaches to studying ethical consumption. And we have already indicated that this book seeks to reconfigure the way in which this phenomenon is theoretically framed. The most productive theoretical framework, developed primarily by Scandinavian-based political scientists, interprets ethical consumption practices as a set of political practices, under the name political consumerism (Micheletti 2003). We discuss this approach further in Chapter 2. One feature of this approach is its reliance on a tradition of grand sociological theory of globalization and modernity. In literature on political consumerism, informed by theories of the risk society and reflexive modernization, the growth of what we are calling ethical consumption is presented as the expression of a broad societal shift away from 'traditional' forms of political participation. From this perspective, ethical consumption appears to be a distinctively novel innovation in political practice, one that stands out against and represents a rupture with norms of 'traditional' politics of parties, elections and interest-group pluralism. This interpretation is, despite its own orientation, prone to be co-opted into academic diagnoses of civic apathy and disengagement. Ethical consumption is just as easily incorporated into narratives of 'neoliberalization', in which it is framed around an analysis of the veiling and unveiling of commodity fetishism in more or less successful ways (e.g., Guthman 2007; Hartwick 2000; Hudson and Hudson 2003; Tormey 2007).

In this book, we seek to avoid the style of grand theorizing associated with narratives of modernity, postmodernity, neoliberalism and advanced liberalism. We seek instead to put to work an analysis of what Michel Foucault called 'modes of problematization'. Foucault suggested that 'the study of problematizations' was 'the way to analyze questions of general import in their historically unique form' (Foucault 1997: 318). Analysing practices in terms of their mode of problematization implies, then, asking 'how and why certain things (behavior, phenomena, processes) became a *problem*' (Foucault 2000: 171). The important point about Foucault's approach from our point of view is that it implies thinking of problematizations not as effects of historical tendencies or conjunctural events, but as indicative of definitive, strategic interventions (ibid.: 172). It is this emphasis that we want to work through in this book, by focusing on the sorts of strategic interventions through which everyday consumption activities have been problematized in specifically ethical registers of global responsibility, and by looking at the practices through which this problematization has been formed and disseminated (cf. Foucault 1986: 11–12).

Our genealogical analysis of the problematizations through which ethical consumption campaigning operates is divided into two parts. Part I, Theorizing Consumption Differently, consists of three chapters setting out the theoretical approach, building on Foucault's notion of modes of problematization, which informs our understanding of consumption. Part II, Doing Consumption Differently, consists of three chapters which work through in a more empirical register the analysis of problematizations and practices of everyday consumption. But these two sections are not sharply divided between the theoretical and empirical. Chapters 2, 3 and 4 critically elaborate on theoretical traditions working in a genealogical vein indebted to Foucault, and these elaborations are informed by our empirical analysis of the political rationalities of ethical consumption campaigning, which suggest that 'strong' hypotheses about neoliberal subjects and advanced liberal governmentality might be in need of some revision. And the more empirical chapters, Chapters 5, 6 and 7, work through the conceptualization of ethical problematization we develop in these earlier chapters.

Chapter 2 develops an argument with the literature on governmentality to build a genealogical conceptualization of the growth of ethical consumption initiatives in terms of strategic interventions which aim to articulate various political programmes (e.g., environmentalism, trade justice, human rights) with the everyday contexts of care-giving and social reproduction.

This argument is informed by a careful analysis of the strategic deployment of the rhetoric of consumer power in ethical consumption campaigning around a double rhetoric of *global responsibility*: simultaneously problematizing people's consumption habits by reference to their distant consequences, and exhorting their potential agency to contribute to transformative projects by exercising choice more responsibly.

Chapter 3 discusses practice-based conceptualizations of consumption. This discussion is informed by empirical analysis of the ways in which ethical consumption campaigns aim not only to provide information to consumers, but also aims to *problematize* everyday practices of consumption by shaping the terms of public debate and by getting people to talk reflexively about their habits and routines.

And in Chapter 4 we further develop this focus on modes of problematization, arguing that a key objective of ethical consumption campaigning is to discursively problematize everyday consumption. The rhetoric of 'choice' and 'responsibility' is central to this discursive problematization of everyday consumption. Building on this argument, we develop a case for understanding the ways in which ethical consumption initiatives deploy information for two purposes: first, to generate *narratives* in both the public sphere and in everyday life, in order to encourage debate about issues of environment, climate change, sweatshops, or trade justice, etc.; and, second, to establish the legitimacy of organizations as *representatives* of popular concerns on these issues.

Chapters 5, 6 and 7 work through in an empirical register the argument about the problematization of consumption through strategic interventions by various actors. Each of these chapters draws on empirical research undertaken in and around the city of Bristol in the south-west of England, involving focus group research investigating the positioning of ordinary people in ethical consumption discourse, interview-based case-study work on local fair trade networks, and ethnographic research on local fair trade campaigning.

In Chapter 5, we analyse talk-data generated in focus groups to gloss ordinary people's understandings of ethical consumption. This chapter shows that people have high levels of awareness of various issues related to the 'ethical' aspects of everyday consumption, and shows too that people bring a range of ethical concerns and competencies to their everyday consumption practices. These range from the relatively personal responsibilities of family life to more public commitments like membership of particular faith communities, political parties, and professional communities. This chapter emphasizes the ways in which people engage critically and sceptically with the demands placed upon them as 'consumers' by campaigns and policies promoting ethical and responsible consumption. Theoretically, this chapter further elaborates on how the discursive problematization of consumption discussed in earlier chapters is negotiated in reasoned forms of talk-in-interaction, in which people consider the degree to which various ethical maxims could and should hold for them.

Chapter 6 further develops the argument that organizations involved in ethical consumption campaigning seek to engage and extend people's existing commitments. It elaborates on how people adopt ethical consumption practices as a supplement to deeper forms of identification, membership and participation. Being involved with fair trade, as a campaigner, shopper or supporter, emerges from this analysis as just one aspect of more extensive practices of 'ethical selving' (Varul 2009). In this chapter, analysis of research on self-identified 'ethical consumers' who are actively involved in local fair trade networks in and around Bristol shows that involvement in ethical consumption is an adjunct to stronger forms of commitment and participation in specific communities of practice. We find little evidence that people adopt ethical consumption activities as an alternative to other forms of civic involvement or public participation. Rather, being an ethical consumer seems to follow from and sustain participation in existing social networks of associational life; and it is these networks which are effectively enrolled en bloc into broader political campaigns.

Finally, in Chapter 7, we pursue further the displacement of 'the consumer' from the centre of analytical attention, by focusing on the dynamics of ethical consumption campaigning directed at transforming urbanized infrastructures of individual and collective consumption. Through a case study of the year–long campaign to have Bristol accredited as a Fairtrade City, we elaborate on how ethical consumption campaigning is just as likely to involve lobbying, negotiating and advising key actors in local authorities, the public sector and private business to adopt 'ethical' supply and procurement policies as it is to focus on efforts to transform aggregate patterns of individual consumer behaviour. The Fairtrade City campaign in Bristol illustrates that campaign organizations operate at different levels to enlist support and transform consumption practices: sometimes they deploy devices that are presented as extending choices to consumers to raise awareness amongst a broad general public and generate media attention; sometimes they engage at an institutional level to change the ways in which consumption is regulated at the level of whole systems of provisioning.

Across these six chapters, we seek to develop three related arguments about the practices and problematizations of ethical consumption.

First, the problematization of consumption through the rhetoric of consumer choice, consumer power and consumer responsibility is a contingent achievement of strategically guided interventions into the public realm and everyday practices. While the rhetoric of consumer agency is certainly prevalent in ethical consumption campaigning, practical interventions seek to engage 'thicker' aspects of people's personal identification (e.g., as parents, as members of faith communities, as professionals); and to change systems of collective provisioning 'behind the backs' of consumers by transforming the design and management of infrastructures of consumption. In short, 'the consumer' is not necessarily the primary target or the main vector of agency in ethical consumption initiatives.

Second, consumption is understood by campaign organizations as a surface of mobilization through which to generate public awareness and enrol potential supporters. This form of mobilization does not necessarily substitute idealized models of consumer agency and market power for other modes of civic participation, associational organization, or collective action. It just as often serves as a pathway for enrolling resources in support of these types of activity.

Third, engagements with consumption-oriented campaigns by ordinary people and by committed ethical consumers alike are guided by attempts to align everyday routines with existing moral and political commitments in order to sustain a degree of personal integrity in an unequal world. Rather than thinking of 'ethical consumers' as individualized utility maximizers or disciplined (or even resisting) subjects of hegemonic ideologies, we argue that understanding the emergence of ethical consumption requires us to take seriously the forms of practical reasoning through which vertical positionings of people as bearers of proliferating global responsibilities are mulled over, acknowledged, or subjected to critical scrutiny.

Part One

Theorizing Consumption Differently

Chapter Two

The Ethical Problematization of 'The Consumer'

2.1 Teleologies of Consumerism and Individualization

This chapter explores how Michel Foucault's discussions of governmentality and ethics can be put to work methodologically to analyse the politicization of everyday consumption. We argue that two aspects of this approach are particularly useful. First, we develop the idea that organizations and networks share rationalities through which they 'problematize' specified areas of social life. And second, we develop the more specific notion of 'ethical problematization', referring to the practices through which people come to take their own activities as requiring moral reflection. In critically developing both of these aspects of Foucault's work on governmentality and ethics, we call into question some taken-for-granted assumptions about the relationship between the politics of consumption and the agency of consumers.

In the United Kingdom since the 1980s, 'the consumer' has been a dominant figure in public debate and policy discourse. In public policy, in particular, rationalities have become oriented towards extending choice and introducing market efficiencies into the delivery of public services (Clarke *et al.* 2007). This apparent triumph of market logic is often legitimized by recourse to a narrative of social and cultural change according to which people's identities and loyalties are no longer defined by reference to work and the labour market, but by what they buy as consumers. For proponents of 'choice', consumer values are an important means of injecting accountability into the public sector. 'Choice' has thereby become a key word in public policy debate in the United Kingdom (Clarke 2010), perhaps even 'the mantra of health, education and pension provision'.[1] The continuing legitimacy of public service provision depends, so the argument goes, on restructuring the public sector in

Globalizing Responsibility: The Political Rationalities of Ethical Consumption by Clive Barnett, Paul Cloke, Nick Clarke and Alice Malpass © 2011 Clive Barnett, Paul Cloke, Nick Clarke and Alice Malpass

line with the expectations of citizen-consumers who now demand the same standards of efficiency, responsiveness, choice and personal flexibility they have become used to in the marketplace (see Clarke *et al.* 2007).

The ubiquity of the choice paradigm should be interpreted as the outcome of a determined effort to recast the balance of responsibility between the state and citizens. What has been dubbed the 'personalization agenda' now 'stretches right across government', encompassing health initiatives and pensions policy.[2] The stated aims of this agenda are to reframe the role of state-led initiatives in terms of empowering individuals to make informed choices, based on information provided by government (e.g., see Pykett 2009). Choice is in turn presented as a means of making service-providers more responsive to the variegated needs of citizens. One can see this individualization of responsibility in a number of fields, extending beyond the realm of the state as such. For example, Hajer and Versteeg (2005) argue that the individualization of health risks has been associated with the burgeoning of socio-cultural practices such as the growth of the fitness industry, self-help publishing and lifestyle media. Likewise, they suggest, in the realm of business, concerns over both health and environment have led to increasing attention being given to the labelling of food products. The discursive individualization of responsibility around various 'risks' or hazards related to personal health and environmental futures leads to considerable faith being invested in the role of information in empowering citizens to pursue their own goals in a way that is conducive to just collective outcomes in markets.

This trend in policy rationalities is often criticized by recourse to the argument that the consumerization of public services undermines the grounds for collective life and public responsibility. On this understanding, the introduction of the logic of choice into public services is just one part of a much more pernicious tendency, whereby the 'triumph of the market' has plummeted us into the 'age of selfishness'.

> The marketisation of everything has made society, and each of us, more competitive. The logic of the market has now become universal, the ideology not just of neoliberals, but of us all, the criterion we use not just about our job or when shopping, but about our innermost selves, and our most intimate relationships. The prophets who announced the market revolution saw it in contestation with the state: in fact, it proved far more insidious than that, eroding the very notion of what it means to be human. The credo of self, inextricably entwined with the gospel of the market, has hijacked the fabric of our lives. We live in an ego-market society. (Jacques 2004)

On this view, consumerism substitutes individualistic self-interest for participation in activities that benefit the public good. This same line of reasoning is often applied to the expansion of ethical consumption, which is criticized for substituting individualized modes of personal responsibility for more collective modes of engagement (e.g., Guthman 2007, 2008). In this type of criticism, consumerism is understood to be the cultural expression of wider rationalities of 'neoliberalism'.

The shared assumption that underwrites the arguments of both proponents and critics of 'consumer choice' is that markets, consumers and choice are all about individualized, materialistic, privatized and self-interest. Both sides hold to a highly idealized model of consumer behaviour, one that assumes that when acting as consumers, people live out the model of the rationalizing utility-maximizer of economic theory:

> Consumers are [. . .] distinctive in the way they make choices (as self-regarding individuals), receive goods (through a series of instrumental, temporary and bilateral relationships with suppliers), and exercise power (passively, through aggregate signaling). (Needham 2003: 14)

Proponents of the market and consumer choice think that people *should* act like this, despite lots of evidence that *they don't*. Critics of the market tend to assume that people *do* act like this, but they think that they *ought not to*, and therefore intone them to act more responsibly.

For critics of consumerism and consumerization, the shared characterization of 'the consumer' as individualized and egoistic allows a rather idealized model of 'citizenship' to be counter-posed to the apparent rise of consumerism. In this model, being a citizen is all about the selfless pursuit of the common good undertaken in collective cooperation with others. According to this criticism of 'neoliberalism', the extension of market relations leads to a depoliticizing individualization, whereby the meanings of citizenship are progressively consumerized (e.g., see Bauman 1999). 'Neoliberalism' is presumed to lead to the intense privatization of experience and identity. According to Elliott and Lemert (2006: 41), the emotional cost of globalization is evident in the extent to which 'people today increasingly suffer from an emotionally pathologizing version of neo-liberalism'. In this type of argument, the privatization and marketization of economic activities is mirrored in processes of subject-formation and identification, which are increasingly focused on self-regulation, self-management and self-sufficiency, for which 'the consumer' is meant to be the paradigm. In the burgeoning field of research on 'neoliberal subjectivities', it is presupposed that programmes of rule unfold by seeking to secure synergies between 'neoliberal' objectives and the motivations and identifications of individuals. If there is such a thing as neoliberalism, then it is assumed that there must also be lots of neoliberal subjects being hailed, more or less successfully, to order (e.g., Bondi 2005; Gökariksel and Mitchell 2005; Guthman and DuPuis 2006; Walkerdine 2005).

By presuming a stark and simple opposition between individualism and collective action, the argument about the rise of 'neoliberal subjectivity' is connected to a claim about the eclipse of the collective dimensions of citizenship. Assuming that consumerism is equivalent to a culture of individualized self-interest, this perspective holds that consumerism is doubly destructive. First, it is a source of harm in the form of environmental degradation, and the reproduction of unequal trade relations and poverty. Second, it also militates against addressing these harms in anything other than a piecemeal fashion by undermining the very possibility of arriving at publicly agreed collective environmental and conservation goals, by privileging an ethos of unfettered individual freedom to pursue the acquisition of material goods. The consumer is therefore an inherently suspect character in these narratives. Schudson summarizes the underlying assumptions of this style of criticism of consumer culture:

> The inferiority of consumer behavior seems to be either that consuming is self-centered whereas political behavior is public regarding or public-oriented, or that consuming, whatever its motives, distracts people from their civic obligations. Either consumption is in itself unvirtuous because it seeks the individual's own pleasures, or its displacement of political activity has unfortunate consequences for the social good. (Schudson 2007: 237)

Debates about consumerism are, then, closely linked to wider arguments concerning the apparent decline in civic activity and political participation. Consumerism has become the central reference point in a wide range of social science arguments which diagnose a broad-based decline of civic life and public participation: consumerism is responsible for the infantalization of public life for Benjamin Barber (2007); the rise of consumer culture leads to the atrophy of civic participation and destruction of social capital for Robert Putnam (2000); consumerism is an expression of a wider onset of cultural narcissism for Christopher Lasch (1979). And these laments stand in a long line of social commentary that sees in modern practices of consumption a destructive culture of 'consumerism' (see Campbell 1990).

Zygmunt Bauman provides a distinctive inflection to this tradition. For him, the rise of consumerism is linked not so much to a decline of public spiritedness, but rather is located within a narrative in which the prescriptions of 'modernity' are replaced by the re-emergence of moral responsibility in 'postmodernity', or what Bauman has come to call 'liquid modernity' (Bauman 2000). Bauman (1993, 1995) presents a narrative in which modernist 'ethics', understood as a field of prescriptive rules, principles and codes regulated by expert legislators, has been displaced by postmodern 'morality', by which he means the primacy of 'responsibility' for 'the other'. Postmodernity therefore marks the freeing up of morality once again, understood as a realm of intuitions embedded in encounters with others, a realm that was previously suppressed beneath ethical codes that substituted rule-following for genuine morality and absolved people from genuine responsibility for their actions. By liberating us from rules, under conditions of postmodernity of liquid modernity we are now exposed to all the burdens and torments that follow from knowing about the consequences of our actions. Responsibility thus emerges as the primary register through which action is undertaken. No longer subordinated to ethical codes, the morality of action now emerges from the rough and ready engagements of life itself. In Bauman's historicist narrative, our actions now appear as 'matters of responsible choice' precisely because the reign of ethical codes based on universal principles and objective certainty is over.

The sting in the tale of this narrative, however, lies in the argument that while 'the state' no longer exercises a monopoly over ethical codes, the supply of ethical rules is now increasingly privatized to the marketplace. Postmodernity transforms morality into the 'tyranny of choice' in which it is not the moral competence of persons that is energized, but rather their 'shopping skills'. For Bauman, 'consumerism' provides the very form of postmodern self-hood, so that responsibility is understood to be a matter of discretionary choice. There is a 'fit', Bauman (2001) argues, between the ethos of consumer culture and how 'postmodernity' requires individuals to produce for themselves the narrative continuities that are no longer provided by society. In this account, Bauman runs together 'individualism' with the act of 'choosing', choosing with 'consuming', and consuming with 'shopping'. Accordingly, we now live in an age in which *Homo consumens* is replacing *Homo politicus* (Bauman 2008). Appealing to stylized facts about the decline of voting, the fall in party membership, and the evacuation of power from the nation-state, Bauman (2007) announces that we now live in a 'liquid-modern society of consumers', a society in which consumption is understood to be a solitary activity, and in which the primary form of collectivity is no longer the group but the 'swarm'.

Bauman locates the pathological account of consumerism within a particularly pessimistic rendition of a long-standing historicist narrative ofsuccessive phases of development, from 'traditional' society to 'modernity', then on to 'postmodernity' and 'liquid modernity'. A more optimistic rendition of this same narrative is provided by theorists of 'reflexive modernization' such as Anthony Giddens and Ulrich Beck (e.g., see Beck, Giddens and Lash 1994). In this line of thought, 'modernity' gives way not so much to 'postmodernity' but to 'globalization', in which tradition loses its hold on people's identities and identifications, and daily life is fully opened up to constant change and flux. In this context, established patterns of identification – class, status and nation – break down and individuals are forced to negotiate among a number of different lifestyle options. Giddens argues that this broad transformation leads to the advent of what he calls *life politics*, which revolves around disputes and struggles 'about how we should live in a world where everything that used to be natural (or traditional) now has in some sense to be chosen, or decided about' (Giddens 1994: 90–91).

Life politics does not, Giddens insists, imply that politics has been reduced to individual choice: 'life politics is about the challenges that face collective humanity, not just about how individuals should take decisions when confronted with many more options than they had before' (ibid.: 92). This hypothesis about life politics is, then, a claim about how collective processes of authoritative decision-making and allocation should be reconceptualized in a context in which individualism itself has been transformed, effectively rendered more individualized than before. For Giddens (1991), the ability to create and maintain biographical narratives becomes the crucial skill for living in a globalized world.

Giddens' account of the narrative elaboration of the self as a distinctively epochal structure of personhood is echoed by Beck (1996, 2006), for whom politics itself has been reinvented by the emergence of global risk society. This social formation is associated with the emergence of novel processes of self-formation: living in a context of heightened insecurity and generalized risk generates a reflexive mode of organizing the self, in which people are faced with constant demands to explain and give accounts of themselves. From this theoretical perspective, the driving force of social change is the process of *individualization* (Beck and Beck-Gernsheim 2001). Individualization refers to the process in which the exercise of agency is effectively freed from traditional social structures; social structures like class cease to be determinant, and people are forced to make decisions about life, work and relationships because the fixed certainties about pathways and norms have, so the story goes, disappeared. It is in this sense that biography, lifestyle and values become a matter of 'choice', in so far as individuals are now obliged to construct for themselves their preferred routes through life. As for Giddens, Beck (1999) sees these processes of individualization as the scene for the emergence of new forms of politics, or 'sub-politics', in which consumerism is identified as a dominant strategy for responding to the constant changes wrought by globalization.

For theorists of 'reflexive modernization' and 'globalization', biographical creativity is placed at the centre of the analysis of how we now live in 'post-traditional' societies. Unlike Bauman, Giddens and Beck do not find in these developments a cause for lament, but identify instead the reconfiguration of traditional understandings of the agents and sites of politics. Both the optimistic and pessimistic version of these claims about the transformations of 'modernity' reproduce a resolutely historicist developmental narrative (see Connell 2007), in which 'globalization' is presented as the real-world process which realizes certain social-theoretical hypotheses (about human agency, reflexive monitoring of conduct, narrative formations of the self) as the cultural dominants of our times. And it is the optimistic narrative of globalized, reflexive modernity that provides the background for the most sophisticated social theoretical account of the contemporary politicization of consumption around 'ethical' concerns.

2.2   Theorizing Consumers as Political Subjects

Once the preserve of academic theories (see Miller 2001a, 2001b; Sassatelli 2007; Slater 1997), the vocabulary of 'consumer culture' and 'consumerism' has now become generally diffused. Both these terms are now 'widely used both to describe a hedonistic postmodern lifestyle and to support a politics of choice' (Trentmann 2006a: 704). Whether as a prelude to a lament or a celebration, the vocabulary of 'consumerism' is routinely equated with choice, self-interest and individualism. The vocabulary of 'consumerism' is, as we have seen in the previous section, bound up with a style of theorizing in which epochs succeed each other: modern by postmodern, materialist by postmaterialist, collectivist values by individualism (ibid.). The idea that the primary attribute of 'the consumer' as a sociological actor is 'choice' should be questioned. So too should the habits of evaluating the extension of choice either as a good in itself; or as eroding shared civic values; or as an effect of the rolling out of neoliberalism or advanced liberalism. All of these formulations reproduce the binaries of consumption, instrumental rationality and autonomy on the one side, and collective values, altruism and public life on the other (see Bevir and Trentmann 2007). This binary moralization of consumption, enshrined in the contrast of 'the consumer' to 'the citizen', deserves to be challenged on various grounds. As Schudson (2007) reminds us, not only can consumer choices express political values and be assembled into political projects, as we argue in subsequent chapters of this book, but consumerism can also enhance democratic, citizenly culture in various ways. The classic argument to this effect is Habermas's (1989) account of the dependence of the formation of a critical public sphere on a consumer culture of books, little magazines and coffee houses. And, as Schudson also reminds us, political choice can be quite a lot like consumer choice: it is bound up with calculations of self-interest, and is hardly the most obvious candidate for an example of public-spirited behaviour.

The primary challenge to the binary framing of *consumer versus citizen* has come through conceptualizations of 'political consumerism' (Micheletti 2003). In fundamental ways, this concept, and related ones such as 'market citizenship' (Root 2007), reproduce the basic understanding that consumer choice is a new medium for political action. The concept of political consumerism has emerged primarily from Scandinavian social science, and builds on sociological theories of globalizing modernity discussed in the last section. Michelle Micheletti (2003: 2) defines political consumerism as 'actions by people who make choices among producers and products with the goal of changing objectionable institutional or market practices'. In this paradigm, political consumerism is understood to be a sociological phenomenon through which the market is being politicized.

In turn, this politicization of the market is taken to be indicative of a trend for people to become disassociated from 'traditional' forms of political life (Micheletti *et al.* 2004: x). For Micheletti, political consumerism is, nevertheless, like citizenship in so far as it involves taking responsibility, solidarity, and the exercise of autonomy. There is a very strong claim made in this paradigm that political consumerism is a distinctively new and original form of political participation (cf. Trentmann 2008), a self-conscious form of political activity which understands itself differently from other forms of participation (Micheletti 2003: xi), and one in which people engage in 'choice situations' to express their political values of justice and fairness (ibid.: 14). The emergence of political consumerism is located within the explanatory frame of historicist sociologies of globalization, in which trends of 'reflexive modernization', the emergence of 'subpolitics', the institutionalization of 'ecological modernization', 'postmodernization' and shifts to 'governance' are presented as the macro-trends explaining epochal shifts in modes of identification and political action. Political consumerism is offered as one aspect of a new conceptualization of politics meant to augment these macro-level concepts.

Micheletti argues that political consumerism represents the emergence of what she calls 'individualized collective action'. Individualized collective action is defined as 'citizen-prompted, citizen-created action involving people taking charge of matters that they themselves deem important in a variety of arenas', which she distinguishes from forms of political engagement 'involving taking part in structured behaviour already in existence and oriented toward the political system per se' (Micheletti 2003: 25). This type of political action flourishes in situations where 'citizens must juggle their lives in situations of unintended consequences, incomplete knowledge, multiple choices and risk-taking' (ibid.). The meaning of 'individualization' in this formulation does not refer to the idea that individual consumer choices can be aggregated up into various 'public choice' functions. Rather, individualized collective action is a reference to Beck's hypothesis of individualization, in which individualization involves the explicit choice of affiliations and identifications. Political consumerism, then, need not be understood as a solitary activity. Understood as 'individualized collective action', it is 'the practice of responsibility-taking for common well-being through the creation of concrete, everyday arenas on the part of citizens alone or together with others to deal with problems that they believe are affecting what they identify as the good life' (ibid.: 25–26). Political consumerism emerges in this paradigm as a form of self-actualizing community formation by newly empowered 'everyday makers' (see Bang and Sorensen 1999), who use their power as consumers to express their own cultural values and social commitments.

The great strength of the theory of political consumerism as a process of 'individualized collective action' lies in its acknowledgement of the ways in which organizations provide people with opportunities to express their commitments and values by purchasing particular products, in turn enabling these organizations to build constituencies of support and influence that become crucial to propelling human rights issues, labour rights issues, or environmental issues into the formal public sphere of political debate and policy making. This acknowledgment of the organized dimension of political consumerism tends, however, to be subordinated to the grander macro-sociological claim that this phenomenon primarily reflects a secular trend in which ordinary people find themselves able and obliged to exercise personal choice in this way. In largely accepting the idea of a consumer-driven phenomenon based on the assertive exercise of choice in the market, the concept of political consumerism tends to mirror back in a positive light the argument that the rise of consumerist forms of 'ethical' action is really only a symptom of a lamentable decline in, and disenchantment with, proper democratic politics. This conceptualization continues to present the politicization of *consumption* as necessarily involving a form of *consumer-*driven politics. It is precisely this taken-for-granted homology between the politics of consumption and the agency of the consumer that we seek to challenge in our analysis.

Jacobsen and Dulsrud (2007) provide a timely critique of the limitations of the macro-sociological conceptualization of political consumerism. They argue that the historicist, developmental style of social theory that this approach draws on leads to a tendency to present the consumer as 'a universally recognizable figure across cultural, historical, and institutional settings' that can be made the target and agent of organized strategies of 'responsibilization' (ibid.: 471). In contrast to the conceptualization that focuses on the agency of consumers and the exercise of conscious choice, they suggest that it is necessary to shift attention 'towards actors and interests that try to impart such responsibilities on consumers, and towards processes intrinsic to consumption that may hamper a widespread approval of such responsibilities' (ibid.: 470). It is exactly this conceptual refocusing that we seek to develop in our analysis of ethical consumption in the chapters that follow.

Jacobsen and Dulsrud (ibid.: 473) suggest that the conceptual focus on consumer agency and attitudes gets in the way of understanding how collective actors frame and mobilize people as 'consumers':

> as for the sovereign active, responsible consumer, there seems to be strong actors within the corporate sector, with governments as well as NGOs that all support the framing of the consumer role and consumption practices in an active direction. An escalation of political consumerism may be congruent with the development of profitable markets, with the de-loading of political and fiscal government responsibilities, and with the power and aims of NGOs. (Ibid.: 475–476)

On this view, political consumerism would be rethought as an effect of the framing and mobilization strategies of organizations, not as an expression of the dynamics of individualization expressed through the mediums of consumer choice. For Jacobsen and Dulsrud, this first shift from focusing on the consumer to focusing on organizations is matched by a second shift which acknowledges that there is a lot more to consumption than the exercise of deliberate consumer choice. The concept of individualized collective action relies on a taken-for-granted understanding of consumption as the activity of consciously choosing from among a range of consumer goods exercised during the act of shopping. This underplays the degree to which a great deal of everyday consumption is routinized, unreflective, and embedded in infrastructures of everyday life. We develop this double displacement in this book – away from the focus on the consumer and on shopping, towards organized assemblages of agency and towards an understanding of consumption as embedded in everyday practices of social reproduction.

The emergence of debates about political consumerism, along with the burgeoning literature on fair trade campaigns, alternative food networks and organized boycotts of 'sweatshop' goods, brings into view the long history of consumer politics. This history clearly indicates that the deepening of commodity consumption does not necessarily lead to a diminution of collective, citizenly practices at all. Far from it, a great deal of modern politics is generated by the deepening of consumption as an infrastructure of everyday life (see Hilton 2005; Trentmann 2008). In contrast to the view that 'the consumer' is an historically novel, generically active figure, the long history of consumer activism suggests that 'the consumer' is framed and mobilized in variable and diverse ways (see Lang and Gabriel 2005). This sense of the contingency of the emergence of 'the consumer' as a political figure is essential to understanding the dynamics of ethical consumption campaigning. Rather than locating this process in the macro-sociological claims of developmentalist globalization theory, it might be better to look at the changing dynamics of non-governmental organizations, social movements and civil society actors in shaping political mobilization. In making this argument, we follow Norris (1998, 2007), who argues that, far from political participation and civic activism being in decline, it is better to think of the reconfiguration of participation and activism. Echoing theorists like Giddens and Beck, she too discerns the emergence of a fluid *politics of choice* which she distinguishes from a *politics of loyalty* based on parties, elections, trade unions. But what is of most value in her account is the more focused account of what drives this emergent politics. First, this new form of activism and participation is distinguished by the *repertoires* used for political expression (e.g., buying or boycotting products, petitioning, demonstrating). Norris calls these 'cause-oriented' repertoires. Second, this new politics of choice is distinguished by the *agencies* and collective organizations who serve as the mediators of engagement and participation.

These tend to be issue-based organizations, and to depend on relatively high levels of expertise. They are certainly distinct from political parties, and are more like advocacy groups than the activist-based organizations which exemplify so-called 'new social movements'.

The focus on *repertoires* and *agencies* draws attention to the extent to which phenomena like 'political consumerism' or ethical consumption are embedded in a broader field of campaigning in which lobbying, activism, advocacy, demonstrating, mobilizing and joining are combined in various ways. The focus on these strategies of organizing and campaigning allows us to see that the 'politics of choice' is not equivalent to a process of 'de-collectivization' at all. But nor should it necessarily be understood in terms of 'individualized collective action', in so far as this concept gives explanatory primacy to the agency of individualizing selves acting as consumers. Thinking of consumer-based forms of expression and mobilization as part of a broad repertoire of political action helps us to see that these are not simply the spontaneous outcome of broad socio-cultural changes of individualization. It is the result of organized activities by strategic actors who are highly attuned to the potentials and pitfalls of consumer-activism. In fact, we would suggest that consumer-oriented activism is *modular*, in the sense that it can be deployed to open up a range of everyday practices to strategic 'ethical' conduct by individuals and organizations (e.g., shopping, investment decisions, and personal banking and pensions), and also because it can be applied to a diverse range of causes (e.g., environmental sustainability, health and safety risks, animal welfare, fair trade, labour conditions and human rights). In both aspects, the application of consumer-oriented repertoires to a broadened range of practices and causes is the outcome of the sustained efforts of a diverse range of organizations and organizational forms, including ethical trading organizations (e.g., Traidcraft, Body Shop, the Fairtrade Foundation); lobby groups (e.g., the Soil Association); trade justice campaign organizations (e.g., Oxfam, Christian Aid); the cooperative movement (e.g., including organizations such as the Co-op in the UK); consumer boycott campaigns (e.g., Anti-Nestlé, Stop Esso); and 'No-Logo' anti-globalization campaigns (e.g., against Nike, Gap, McDonald's).

In short, the growth of ethical consumption campaigning takes place when there is a concerted effort by organizations and institutions. This is not to deny that there are all sorts of latent values involved in being a 'consumer' that these campaigns are able to tap into. But the narrative of individualization tends to overestimate the degree of discretion people exercise over much of their everyday consumption activity. In Chapter 5, we use focus group evidence from a range of social areas in Bristol to gain some insight into what Sayer (2004) calls 'lay normativity' – the everyday motives, norms and values that shape people's conduct and behaviour. Not surprisingly, we found that people are concerned about value for money, which may seem like a rather self-interested, perhaps individualistic concern. But this concern with value for money is embedded in much broader concerns that people have about, for example, what to put in their kids' packed lunches, or about the health impacts on themselves and their loved-ones of different sorts of foods that they buy. Ordinary consumption is already shaped in all sorts of ways by values of caring for other people, and, sometimes, by quite explicit moral values drawn, for example, from the faith communities or ethnic groups to which they belong. Two things follow from this. First, there is a diverse range of values and commitments that guide people's routine shopping, investment and consumption practices, ranging from personal health and local community issues to environmental values, faith-based commitments and concerns over global poverty. The second point is that a great deal of the 'consuming' people do is not undertaken by them as 'consumers' at all, but is embedded in other sorts of practices where they are enacting other identities. Daniel Miller has argued that rather than thinking of commodity consumption as a field of selfish individualism, it is better to think that we appropriate the world of consumer goods 'in order to enhance rather than detract from our devotion to other people' (Miller 2001c: 231). Appreciating the ways in which consumption is embedded in practices of sociability, generosity and care helps us recognize just how much 'shopping is directed towards others, particularly family members, and how far it is guided by moral sentiments towards them and about how to live. Far from being individualistic, self-indulgent, and narcissistic, much shopping is based on relationships, indeed on love. It often involves considerable thoughtfulness about the particular desires and needs of others, though it may also reflect the aspirations which the shopper has for them, thereby functioning as a way of influencing them' (Sayer 2003: 353). Furthermore, once we recognize these relational dimensions of everyday consumption, it also becomes evident that a great deal of 'consumer' behaviour – shopping being the obvious example – is not necessarily best thought of as an arena of 'choice', but is rather constrained and shaped by all sorts of obligations and responsibilities (Miller 1998).

In section 2.1, we looked at traditions of sociological thought which see the rise of consumerism, understood as a vector of choice, as the teleological outcome of the dynamic of modernization and globalization. In this section, we have discussed the ways in which this tradition has provided the grounds for an innovative conceptualization of 'political consumerism'. Both of the general theories discussed in the first section and the specific analysis of political consumerism discussed in this section fit what Dean (2007: 61) labels the culture-governance thesis, sharing as they do the claim that governance increasingly operates through 'ethical culture or cultivation of the individual':

> This is the view that rule in contemporary liberal democracies increasingly operates through capacities for self-government and thus needs to act upon, reform and utilize individual and collective conduct so that it might be amenable to such rule. (Ibid.)

This is the background understanding against which globalization is understood to generate the constant calling into question of received identities, and the reframing of individual identity as a matter of choice. Bauman's pessimistic vision of the rise of *Homo consumens* shares in this framework just as much as Giddens' or Beck's optimistic embrace of reflexivity and individualization. While there is much that is suggestive in these types of analysis, we want to mark our distance from this style of teleological, developmentalist social theory. Rather than thinking of the emergence of consumerist repertoires and new agents of ethical consumption as reflective of secular sociological trends, we follow Dean in locating these processes in relation to situated deployments of various forms of power 'which divide populations and seek to fabricate specific forms of individuality, to the production of self-evident truths for public policy and governmental practices, and to the normalization of particular ways of life embedded in a particular social and political order' (ibid.: 78). This genealogical perspective follows from Dean's advocacy of an alternative approach to the contemporary dynamics of individualization and governance through ethics, one that builds on the work of Foucault on governmentality and modern political rationality. In the rest of this chapter we develop our own critical perspective on the utility of this alternative theoretical tradition for the analysis of the politicization of consumption through the growth of ethical consumption campaigning. In so doing, we argue that the various 'powers' referred to above need to be thought of as extending beyond the usual institutions discussed in either theories of 'governance' (the state and private sector) or theories of 'governmentality' (the state and pacified civil society), to include a broad range of social movements actors involved in innovative forms of contestatory political action. We seek, in short, to contribute to the analysis of 'alternative governmentalities' of the sort Rose (1999: 274–284) begins to sketch, without fully developing, at the close of his influential account of advanced liberal governmentality, *Powers of Freedom*. We do so by fleshing out a genealogical analysis of ethical consumption, one that understands the historicity of this phenomenon as an effect of the contingent combination of particular 'modes of problematization' (Foucault 1997: 318).

2.3 The Responsibilization of the Consumer

We have seen in the previous two sections that accounts of consumerism and the consumer are often couched in an explicitly historicist register. Consumerism is always on the rise, eclipsing other, supposedly more virtuous, forms of social interaction. This historicism is at the root of the moralization of consumption that is endemic in social science literatures. It tends to erase from view the variability of consumer subjectivities (Trentmann 2005). And it ignores the degree to which this variability is dependent on the active facilitation of consumer subjectivities by strategic actors, including the state, corporations and companies, and non-governmental organizations. In order to restore this double emphasis, in this section we critically develop some insights from genealogical approaches to conceptualizing 'the consumer'.

The most influential genealogical approach to understanding the contemporary prevalence of the figure of 'the consumer' is work informed by Foucault's ideas on 'governmentality' – what we will refer to here as 'governmentality theory' (see Dean 1999). This work sees the rise of 'the consumer' as just one effect of a thoroughgoing transformation in the political rationalities governing relationships between states, citizens and markets. Under so-called 'advanced liberal' styles of government, the concept of the citizen is transformed from one based on a notion of a subject with entitlement rights against a social state, to a 'responsibilized' subject modelled on the consumer who activates personal preferences in the marketplace (Larner 1997). The key point of this account is that the rise of the consumerized-citizen does not just follow automatically from shifts in the social relations of production, distribution and consumption, or from general trends of modernization or secularization. It is, rather, an active achievement brought about by many different actors and 'marked by the proliferation of new apparatuses, devices and mechanisms for the government of conduct and forms of life' (Rose 1999: 164). From this perspective, the consumer is 'mobilized' in different ways by various actors who make it possible for people to act as consumers – that is, as choosing subjects (Miller and Rose 1997; Binkley, 2006). In governmentality theory, 'advanced liberal' rationalities position individuals in an array of 'ethical' projects, where ethical refers to 'projects in which individuals are asked to work on themselves and their conduct and to transform themselves with the help of experts, training and services' (Dean 2007: 63; see also McDonald *et al.* 2007).

Governmentality theory focuses on the diversity of agents, knowledges and technologies involved in working up the 'consumer' as a surface of government. The relevance of this approach to understanding the emergence of ethical consumption has been demonstrated by Lockie's (2002) analysis of the growth in Australia's organic food sector (see also Lockie and Goodman 2006). This is explained not as a response to consumer demand but rather by reference to the active dissemination of discourses of ethical responsibility by intermediaries including supermarket retailers, nutritionists and market researchers. So mobilizing the 'ethical consumer' has a double-sided aspect to it. On the one hand, it involves organizations making practical and narrative resources available to people to enable them to act as 'responsible' subjects not only in relation to their own circumscribed criteria of utility but also in relation to broader social and environmental 'responsibilities'. On the other hand, it involves organizations making a collective of 'consumers' knowable through market research, surveys and other technologies in order to speak in their name in policy arenas and the public realm.

The genealogical approach we are recommending here underscores the historically contingent relationship between the problematization of consumption on the one hand, and the mobilization of subjects as consumers on the other (Barnett *et al.* 2005). It has become a staple of analysis of market-oriented 'neoliberal' governance and 'advanced liberal' governmentality to examine the ways in which various policy initiatives seek to govern the capricious virtue of 'free individuals' (Newman 2006). What is notable about contemporary ethical consumption campaigning in this respect is that it does not necessarily mobilize a sense of the consumer as an economistic self-interested utility maximizer, as is often assumed in theories of neoliberalization. Rather, the consumer is invoked as the bearer of a variety of responsibilities, both for their own self and for dispersed others. Likewise, policy interventions are increasingly redefined in terms of a shared logic of 'responsibility', in which the greater freedom ascribed to individuals as consumers in markets for goods and services needs to be balanced by efforts to instil in them greater concern to look out for their own good (in terms of health, diet, education or security) and for various collective goods as well (such as environmental conservation, global poverty, climate change).

In seeking to account for the proliferation of discourses of consumerism across various fields of 'government', Nikolas Rose has argued that advanced liberalism involves a thoroughgoing reordering of the ways in which political rule is exercised. This reordering is expressed in a shift from governing through society to governing through individuals' capacities for self-realization. For Rose, political rationalities of advanced liberalism are characterized by 'individuals and pluralities shaped not by the citizen-forming devices of church, school and public broadcasting, but by commercial consumption regimes and the politics of lifestyle, the individual identified by allegiance with one of a plurality of cultural communities' (Rose 1999: 46). This shift has led to the suturing together of the status of citizenship with the diverse roles of being a consumer:

> Advanced liberal forms of government thus rest, in new ways, upon the activation of the power of the citizen. Citizenship is no longer primarily realized in a relation with the state, or in a single 'public sphere', but in a variety of private, corporate and quasi-public practices from working to shopping. (Ibid.: 166)

The new political subjectivities emergent from the reframing of citizenship in this way are, Rose continues, linked to new practices of identity-formation:

> These fuse the aim of manufacturers to sell products and increase market share with the identity experiments of consumers. They are mediated by highly developed techniques of market research and finely calibrated attempts to segment and target specific consumer markets. Advertising images and television

programmes interpenetrate in the promulgation of images of lifestyle, narratives of identity choice and the highlighting of the ethical aspects of adopting one or other way of conducting one's life. (Ibid.: 178)

And all this amounts to a 'new habitat of subjectification', one characterized by 'the belief that individuals can shape an autonomous identity for themselves through choices in taste, music, goods, styles and habits' (ibid.). Developing this type of analysis, Shamir (2008) goes so far as to argue that we live in an age of 'responsibilization': the economization of the political sphere under generalized neoliberal governance, so this argument goes, generates a dialectical moralization of economic action. 'Responsibilization' is understood here as a technique of governance 'premised on the construction of moral agency as the necessary ontological condition for ensuring entrepreneurial disposition in the case of individuals and socio-moral authority in the case of institutions' (ibid.: 7).

Governmentality theory tends to locate any trend of 'responsibilization' within a narrative of advanced liberal governmentality, understood as marking an epochal shift from state to market provision and regulation. It largely ignores the degree to which patterns of responsibilization are an outcome of political contestation and struggle rather than having a single dialectical logic. The 'responsibilization' of the social field is not only the work of states, governments and policy makers. Key actors here include capital (in the form of corporate social responsibility initiatives, for example) and a whole range of non-state actors such as charities, NGOs and campaign groups. There is no single logic unfolding here, but rather a set of overlapping programmes and interests (Larner *et al.* 2007). If, then, it makes sense to identify 'responsibility' as a master frame of contemporary public culture, then it also needs to be acknowledged that this is 'a cultural model of social practice that has first emerged from the dynamics of the discourse generated by the new social movements, the state and business or industry in the course of antagonistically addressing the issue of risk, and then has become established beyond them at the macro-level' (Strydom 1999: 75). The sense that responsibilization is a process that involves diverse actors, pursuing diverse objectives, provides a better entry point to understanding the ways in which the politicization of consumption has come to be framed around discourses of consumer responsibility than the singular emphasis on a shift from liberal to advanced liberal political rationalities found in governmentality theory.

The aspect of the Foucauldian narrative of advanced liberalism that is valuable for our analysis is the argument that the contemporary suturing together of citizen and consumer involves not so much a straightforward diminution of citizenship, but more the emergence of a more 'ethical' mode of citizenship practice and identity-formation. 'Ethical', in governmentality theory, refers to the active shaping of lives in relation to individuals' own sense of fulfilment rather than by reference to models of citizenship, in which obligation and prescription are the dominant registers of subject-formation (Rose 1999: 178–179). This perspective leads us to propose that the contemporary politics of *consumption* is articulated through various modes of 'ethical problematization', whereby people are expected to treat their conduct as *consumers* as subject to all sorts of moral injunctions. It is in this sense that the claim that 'mobilizing the consumer' has become a key aspect of current rationalities of political rule should be interpreted (cf. Miller and Rose 1997). Governmentality theory helps us to avoid the trap of simply bemoaning the death of citizenship and the eclipse of public virtue in the face of a capacious consumerism. As Rose (1999) acknowledges, governing through citizenship is inherently risky: the bearers of citizenship status are always likely to reinterpret ascribed rights and make counter-claims against those actors to whose authority they are subjected. This indeterminacy is only likely to be exacerbated when 'citizen' gets articulated with 'consumer' as a prevalent register of subject-formation. We can expect, then, that consumption is a field of intense contestation between competing rationalities of the free market, of rights and participation, and both the hedonistic and caring dimensions of everyday consumer practice. If, following Trentmann (2006b), we think of the relationship between systems of commodity provisioning and consumer identity as historically contingent, then we need to acknowledge that commodity consumption does not necessarily produce a self-understanding of oneself as a consumer, and also that the politics of consumption is not necessarily always articulated through forms of consumer politics. As Gabriel and Lang (1995) have argued, the consumer is a historically variable figure mobilized by different interests at different points – sometimes as a chooser, or as a communicator, or as a victim, or as a citizen, or as an identity-seeker, or as an activist. This implies that the politics of consumption is analytically distinct from, and of wider scope than, consumer politics per se.

The starting point for our genealogical analysis of ethical consumption is, then, the observation that the history of consumer activism belies the assumption made in so much social theory that consumerism is a vehicle for narrow self-interested egoism. This history illustrates the degree to which the rise of consumerism has long been associated with innovative ways of expressing other-regarding concerns and solidarities (Hilton 2005, 2007a; Sassatelli 2006). Consumer activism, which focuses on the mobilization of people to secure various rights as consumers in the marketplace (quality, safety, fair pricing and so on), is distinct from the use by various movements of the repertoires of consumerism as a means through which to mobilize support or attention for causes that extend the arena of consumer rights itself. The labour movement, cooperative societies, trade justice campaigns, the peace movement and gay and lesbian rights campaigns have all constructed 'the consumer' as a subject-position through which people can exercise broader rights and obligations as citizens. Contemporary ethical consumption builds on the solidarity concerns of the latter sort of movements, but seeks to embed these in everyday concerns with the quality of goods and services consumed in homes and workplaces.

In the next section we step back from consumption per se, and examine how the relationship between technologies of government and practices of subject-formation is theorized from a governmentality theory perspective. We do so because, while acknowledging that the turn to concepts of governmentality injects a much needed sense of difference into narratives of neoliberalization (see Larner 2003), this observed shift still derives from an assumption that the dynamic of political processes is initiated from the top down. There are two problems with this assumption. First, the analytics of governmentality has come to be a resolutely 'policy-centric' approach to understanding socio-cultural change. Everyday life and social relations are reduced to residual effects of initiatives emanating from diverse, but nevertheless coherent, concentrations of authority. 'The social' is defined as a fundamentally *reactive* field, rather than one from where dynamics of socio-cultural change might actually emerge. This misconstrues the diverse modes of political action that might be pursued through practices of consumption. Second, the way in which the immanence of contestation to governmental rationalities is theorized means that this approach cannot, in the final analysis, actually account for the 'success' or 'failure' of programmes of rule either empirically or conceptually. These two problems suggest that 'political rationalities' should be ascribed a more modest influence in shaping the trajectories of social change.

2.4   What Type of Subject Is 'The Consumer'?

One common criticism of governmentality theory is that this approach presumes that the subject-effects inscribed in rationalities and technologies of government are automatically realized in practice. In fact, theorists of advanced liberalism are clear that they do not suppose that governmental rationalities automatically determine subjectivities:

> The forms of identity promoted and presupposed by various practices and programmes of government should not be confused with a *real* subject, subjectivity or subject position, i.e. with a subject that is the endpoint or terminal of these practices and constituted through them. Regimes of government do not *determine* forms of subjectivity. They elicit, promote, facilitate, foster, and attribute various capacities, qualities and statuses to particular agents. They are successful to the extent that these agents come to experience themselves through such capacities (e.g. of rational decision-making), qualities (e.g. as having a sexuality) and statuses (e.g. as being an active citizen). (Dean 1999: 32)

This is the characteristic vocabulary of governmentality theory, replete with terms such as 'elicit', 'promote', 'foster', 'attract', 'guide', 'encourage' and so on. While Dean implicitly leaves open a space for the analysis of processes of *identification* (ibid.), nevertheless even this avowal of non-determinism tends to pre-construct any such analysis in particular ways. This is illustrated by Lemke's account of neoliberalism:

> The key feature of the neo-liberal rationality is the congruence it endeavours to achieve between a responsible and moral individual and an economic-rational actor. It aspires to construct prudent subjects whose moral quality is based on the fact that they rationally assess the costs and benefits of a certain sort as opposed to other alternative acts. (Lemke 2001: 201)

The emphasis here is on a rationality that '*endeavours*' and '*aspires*' to bring about certain subject-effects. Taking theorists of governmentality at their word, they only claim to be able to identify emergent rationalities and associated technologies of governing (Rose *et al.* 2006). The problem with governmentality theory, when it comes to thinking about subject-formation, is not degree of closure around intended subject-effects. This approach leaves plenty of room for contestation and variable identification. The main problem with this approach is the overwhelmingly *strategic conceptualization of action and interaction* through which the processes of identification and contestation are theorized.

To illustrate what we mean by this, consider Dean's paradigmatic definition of 'government':

> Government is any more or less calculated and rational activity, undertaken by a multiplicity of authorities and agencies, employing a variety of techniques and forms of knowledge, that seeks to shape conduct by working through our desires, aspirations, interests and beliefs, for definite but shifting ends and with a diverse set of relatively unpredictable consequences, effects and outcomes. (Dean 1999: 11)

Three aspects of this definition are noteworthy. First, the emphasis on the multiplicity, variety and diversity of ends, means and outcomes of government explains the wide range of fields to which ideas of 'governmentality' can be applied as a tool of analysis. Second, Dean claims that government in this Foucauldian sense works through specifically 'subjective' modalities of desire, aspiration, interest and belief. Third, this definition presents government as a particular type of action: it is calculated, rationalized; it seeks to shape conduct; and it is oriented by certain ends.

The third aspect of Dean's definition is the source of the fundamental theoretical limitation of theories of governmentality: they imagine all forms of action as primarily *strategic* action. Questions of ethics, freedom and liberty are understood by reference to a quite specific notion of games, wherein the aim of the contest is 'structuring the possible field of action of others' (Lemke 2002). This emphasis on power as a form of strategic action overdetermines the emphasis on multiplicity, variety and diversity that one finds in this line of thought. It is here that the conceptual and empirical limitations of theories of advanced liberalism need to be located.

The analytics of governmentality is meant to be a distinctive approach to understanding political power. Adopting this perspective means that one should 'start by asking what authorities of various sorts wanted to happen, in relation to problems defined how, in pursuit of what objectives, through what strategies and techniques' (Rose 1999: 20). This sounds like a modest methodological claim. Looking at what authorities of various sorts *wanted to happen* can be a productive entry point into researching phenomena such as administration, organizations and policy. But governmentality theory is presented as more than simply a methodological approach. This starting point ends up covering the entire field of what is understood to be legitimate social science analysis. Governmentality theorists develop a fully loaded theory of the exercise of political power from this starting point, one which ends up ascribing a considerable degree of causal significance to the play between competing strategic rationalities. This approach holds that ways of visualizing phenomena have a high degree of effectivity in actually bringing those phenomena into existence, as so many 'contingent lash-ups of thought and action'.

Governmentality theory tends to mistake a methodological entry point – investigating what authorities wanted to happen, in relation to problems, in pursuit of objectives, through strategies and techniques – with the primary causal dynamic of social practices. The plans and programmes which disclose political rationalities are explicit and easily comprehended. The analytics of governmentality does not acknowledge the degree to which the rationalities that govern strategic interactions are not the pre-existing properties of the different actors involved, but are an emergent dimension of ongoing interaction itself (Bridge 2005: 106–107). Acknowledging the em*ergent rationalities* of interaction implies that the concatenation of strategic interests often leads to various forms of cooperative behaviour – bargaining, helping, compromising and self-binding (Elster 1989). And this requires a shift away from a wholly strategic conception of action and interaction, since this conception is not able to acknowledge the capacities for communicatively mediated, normatively oriented interaction through which such emergent cooperative rationalities develop (see Boltanski and Thévenot 2000; Stark 2009).

The conceptual adjustment we are recommending revolves around a dissonance between Foucault's later work on governmentality and his work on ethics. There is a double gesture in this work on political rationalities and the work on sexual practice, ethics and technologies of the self (see McCarthy 1993: 63–75). First, in the work on governmentality, power is explicitly theorized in ways that distinguish it from violence and domination, refining the notion of the productive qualities of power relations (Foucault 2000: 326–348). Second, in the work on ethics, the subject is analysed not only in terms of subjectification, but also in terms of the care of the self (Foucault 1985). Nonetheless, Foucault's own analytics of governmentality continues to privilege a conception of social interaction in terms of 'strategic games of liberty' 'in which some try to control the conduct of others, who in turn try to turn to avoid allowing their conduct to be controlled or try to control the conduct of the others' (Foucault 1997: 299). The 'action on the action of others' that defines governmentality as a distinctive rationality of rule continues to be theorized by reference to actors' efforts to determine the conduct of others, to realize their own ends through the enrolment of the strategic capacities of other actors. The fundamental commitment to a strategic model of power remains intact through these developments in Foucault's later work: 'Foucault's final ontology tends to equate social interaction with strategic interaction' (McCarthy 1993: 63). So while Foucault certainly loosened up the sense of disciplinary power in his account of governmentality, he did so only to pluralize the underlying sense of strategic action as the medium through which different actors are bound together in joint projects. He differentiated between three senses of strategic relations: a fairly neutral understanding of means–ends relations; a sense of taking advantage of others; and a sense of obtaining victory in struggle (Foucault 2000: 346). But he endorsed the idea that these three senses covered the whole field of power relations, where strategy was understood as 'the choice of winning solutions' (ibid.) in situations of confrontation or competition.

The question that concerns us here is whether the twin moves that inaugurate the thematic of governmentality in Foucault's work – specifying the productivity of power as distinct from domination, and focusing on practices of the self as distinct from subjection – can be combined into an analytics that is capable of freeing social action from its conceptual residualization in so much work on advanced liberal governmentality and hegemonic neoliberalism alike (see Barnett 2009). In order to recuperate the generative dynamics of the social, it is hardly enough to affirm the creative, inventive, resistant responses of subjects positioned by overbearing discourses and technologies of subjection. Nor is it adequate to ascribe this capacity to a thinly theorized account of the vitalism of 'life itself' generating various efforts at 'de-subjectification'. Governmentality theory, just like theories of hegemony, is far too averse to acknowledging the degree to which normativity inhabits the social world in various ordinarily communicative registers (Boltanski and Thévenot 2006), an issue we return to in detail in Chapter 5. Without this acknowledgement, any analysis of the *identifications* through which political rationalities extend themselves, of the sort hinted at by Dean (1999: 32–33), will remain constrained within a determinism/resistance problematic.

What is most valuable in governmentality theory is the attention it invites us to pay to *how* programmes of rule seek to bring about their objectives. In the next section, we look at just this question, and it leads us to revise one of the basic assumptions of governmentality theory. We question whether attempts to govern consumption in line with principles of sustainability, fair trade and global trade justice do in fact aim to govern conduct through operating directly on people's identifications – by 'getting at' their desires, motivations and beliefs. Examining the deployment of technologies of calculation in this field of campaigning and policy reveals a form of rationality marked by a much more convoluted engagement with people's identifications. In turn, we suggest that the primary subjects targeted by such initiatives are just as likely to be specific professionals and experts, not just generic 'consumers'. In section 2.6, we then turn to the question of how to conceptualize consumption as a realm in which people's conduct is governed 'at a distance' through the operation of norms. We develop the argument that theories of governmentality cannot recognize the mechanisms of any such practice in so far as they are characterized by a recurrent tendency to collapse the distinction between governing *action* and governing *subjectivity*.

2.5   Does Governing Consumption Involve Governing the Consumer?

The contemporary politics of consumption touches on central issues in debates about neoliberalism and advanced liberalism: the relationships between states and markets, citizens and consumers; and ascriptions of responsibility amongst a range of public, commercial, societal and individual actors. Ethical consumption campaigns problematize current patterns of commodity consumption on the grounds that they are environmentally destructive, help to reproduce global inequality, and are complicit in human rights abuses (Harrison *et al.* 2005). These campaigns are connected to, but not identical with, mainstream policy debates about sustainable consumption. In policy debates around climate change, or environmental pollution, or global poverty, it is often the default assumption that changing *consumption* requires finding ways to make *consumers* change the way they exercise choice. It is also widely assumed in academic debates that rising levels of unsustainable consumption are maintained by a culture of consumerism, in which people's identities are thoroughly invested in their status as consumers. This latter position provides a degree of critical purchase on pro-market discourses only by compounding the underlying assumption of those discourses, namely that people's identities are in fact more and more patterned by their roles as consumers. We need to call into question the assumption that governing consumption necessarily involves governing people's identifications as 'consumers'.

One of the most productive areas of research informed by theories of governmentality is research on the diversity of 'calculative practices' associated with the rise and transformation of the modern state. In his discussion of the historical significance of accounting and accountancy, Peter Miller (2001: 379) suggests that calculative practices 'alter the capacities of agents, organizations and the connections among them', and that they 'enable new ways of acting upon and influencing the actions of individuals'. But just what sort of power is this? How exactly do calculative practices enable this 'acting upon' and 'influencing'? In just what sense can one think of calculative practices as aiding in what Miller and O'Leary (1987) have called the 'construction of governable persons'?

One helpful way of understanding the relationship between calculative practices and subject-formation is to turn to Ian Hacking's (2002) account of how the 'avalanche of numbers' through which modern states have categorized people since the nineteenth century have generated new ways of 'making up people'. Systems for counting populations depend on forming categories around people, and these categories in turn help form those people in so far as they mobilize around those classificatory identities. Hacking allows a much greater degree of proactive autonomous pressure from 'below' than other Foucauldian-inflected analyses. This is related to his emphasis on the variable interactions between classifications of people and the people classified, an understanding that he dubs 'dynamic nominalism' (Hacking 1999). He argues that the classification of people is shaped by various 'looping effects', through which any adjustment that people make to their own conduct as a result of being classified in new ways renders those classifications false, thereby requiring an adjustment, which generates further adjustments, and so on and so on.

We can see an example of this process in contemporary ethical consumption campaigning. Calculative technologies such as surveys and polls are important aspects of campaigning around ethical, green and sustainable consumerism in the UK. Between 2001 and 2006, the Co-operative Bank in partnership with London-based think-tank the New Economics Foundation (NEF) has produced *The Ethical Purchasing Index* (EPI).[3] Using criteria that include fair trade, vegetarianism, organic foods, green household goods, buying locally and responsible tourism, this annual report tracks the growth and *potential size* of the market in ethically produced and traded goods and services. These survey data are deployed in a wide range of public arenas. The EPI was used to generate widespread newspaper and broadcast media coverage when published, and is part of a rolling stream of such survey research. One set of addressees for this type of survey are, then, members of the general public. It is one part of a broad strategy of raising awareness of a range of 'ethical' consumption issues.

However, these calculative technologies are also deployed to lobby government and businesses to extend support for ethical consumer initiatives. It would be wrong, then, to assume that the primary objective of these exercises is to change the consumer's own conscious behaviour directly. Rather, the intention is also focused on enhancing the supply of 'ethical' commodities through exerting pressure on retailers and suppliers – and, just as importantly, to represent consumer choice as a *political* preference in the lobbying of policy makers, regulators and corporations.

Exercises in enumerating ethical consumption such as the EPI work a lot like Hacking's dynamic looping effects. Campaigns to raise awareness and encourage people to exercise consumer choice 'ethically' lead to a disparate set of purchasing acts that are classified, counted and represented in new ways in the effort to alter retailing practices, and procurement and supply policies. In so far as these alterations take place, they in turn facilitate further acts of 'ethical' purchasing by anonymous consumers, which are classified and counted again in new rounds of surveying. The calculative practices involved here act on people's actions in a very indirect way. They certainly seek to structure possibilities of action, but hardly by endeavouring to transform people's subjectivities. The deployment of survey data in ethical consumption campaigning indicates that, as a mode of classifying people, 'the consumer' might be best thought of as a relatively 'indifferent kind', in contrast to the sorts of 'interactive kinds' where there is a strong degree of fit between classification and classified which is mediated by explicit identification with categories (see Hacking 1999). Being classified as a 'consumer', ethical or otherwise, is only rarely taken up as a strong point of personal identification. This is not to say that the classification does not have effects, only that this is not necessarily mediated through strong interpellative subject-effects.

There are two points we want to draw out from this example. First, there is no need to suppose that efforts to shape conduct aim to bring about strong subject-effects on individuals who identify themselves as 'ethical' consumers. Looking at *how* calculative technologies enable new ways of acting on individuals suggests that the aim is not to generate specific subjective identifications, but is rather to enable various sorts of acts. Second, the agents likely to be the subjects of governmental rationalities are not necessarily myriad, dispersed, anonymous 'consumers'. They are just as likely to be various professionals and experts working in key areas of decision making, such as procurement, purchasing or management (Newman 2005). Crucially, professionals in these fields are not just the passive agents of anonymous governmental rationalities; they are just as often the agents of such interventions. We discuss this in more detail in Chapter 7, where we examine the dynamics involved in the campaign to have Bristol certified as a Fairtrade City by the Fairtrade Foundation (FTF). In Chapter 7, we show that Bristol's Fairtrade City campaign involved the creative response by local authority professionals to an initiative from non-governmental organizations (the FTF) in which the only tangible stake was gaining the kudos of being certified as a Fairtrade City. The example of Fairtrade City campaigns indicates that seeking to govern consumption does not necessarily require direct interventions to regulate individual consumer choice at all. Innovative efforts aimed at transforming the collective infrastructures of individual consumption are just as likely to be the focus of programmatic interventions. This example also indicates the diverse range of subjects targeted by such programmes. A fundamental determinant of the path that campaigns of this sort take in different places will be the degree to which different actors are able to be 'empowered' by such initiatives. Responding to central government sustainability initiatives does not involve these actors simply applying a set of rules and principles handed down from above. For example, Morgan and Morley (2004) suggest that governing complex systems such as food chains requires an appreciation of the degree to which different professional actors are able to *interpret* different rules, regulations and policy agendas. Their point, one borne out by our own example of the Bristol FTC campaign, is that these professional actors are not mere bearers of a 'neoliberal' agenda, but that they have significant discretion for creative, proactive interventions in the networks into which they are placed (see also Newman 2005). And governmentality theory has great difficulty accounting for this capacity for intra-organizational innovation (Newton 1998).

In this section, we have made two related arguments. First, the key site of interventions into consumption are just as often the infrastructures of consumer choice as they are direct injunctions to individual consumers to change their behaviour. Our second point is that these sorts of interventions aim to reshape the *actions* of consumers, certainly, but that they might well be relatively indifferent to the subjective motivations of individual consumers. In the later chapters of this book, by asking empirical questions about how the rationalities and technologies of governing consumption in explicitly 'ethical' or 'sustainable' ways actually work, we show that there is a need to be more circumspect about the general applicability of grand claims about the individualization of contemporary governance. In the next section, we want to pick up the point made above about the importance that the *interpretation* of rules and regulations plays in shaping the networks through which 'governmental' interventions play out. The claim that rules and regulations only work through the medium of interpretation cuts to the conceptual and empirical aporia around which Foucauldian accounts of advanced liberalism, biopolitics and neoliberal governmentality turn. These accounts rely on the idea that modern modes of rule operate through the distinctive action of norms (as distinct from juridical interdiction; see Ewald 1991). As Hammersley (2003: 753) observes, norms 'do not include instructions for their own interpretation', since every application of a norm requires a capacity for interpretation that is not rule-governed, 'in the sense that it involves identifying a situation as being of a kind that is relevant to a norm, or to one norm rather than another, and recognizing what the implications of the norm are for action in that situation' (ibid.: 754). Following a line of thought that goes back to Garfinkel and is indebted to a phenomenological view of subjectivity, being subjected to a norm depends on the capacity for practical reasoning: performing in a norm-governed environment involves normative capacities of giving reasons, deliberating and providing accounts (see Dreyfus 2007a, 2007b; McDowell 2007a, 2007b). Poststructuralist perspectives which reduce the action of norms to the entrainment of muted bodies into routinized and repetitious patterns of conduct close off our understanding of this aspect of the normativity of norm-governed action; so too does any account that reduces action to a purely strategic encounter devoid of any communicative aspect at all.

2.6   The Ethical Problematization of the Consumer

Foucault's strategic understanding of power relations and technologies of the self underlies the distinctive questions opened up by governmentality theory. In this paradigm, subjectivity is no longer reduced to the operation of mechanisms of subjection. With a nod to Habermas, Foucault proposed that the idea of 'technologies of the self' was a fourth type of technology through which human affairs could be understood, the others being technologies of production, communication and power (Foucault 1997: 224–225). Technologies of the self are means by which to determine 'the conduct of individuals and submit them to certain ends' (ibid.: 225). The analytics of governmentality inserts itself at the intersection between technologies of domination and technologies of the self. Any consideration of communicative action, which one might suppose could be important in mediating this relationship, is dismissed. Foucault (ibid.: 298) suggests that his difference with Habermas lies in the fact that the latter gave 'communicative relations' a utopian position. This might well be true, but Foucault commits the reverse error. His scruples about providing normative *foundations* for critique lead to an inability to acknowledge the ordinary ways in which normativity might inhabit human affairs in and through communicative action. These scruples have now ossified into a pattern of mutually reinforcing theoretical criticism, for which Foucault and Habermas have come to stand as figures. (For reconsiderations of this pattern of theoretical exegesis, see Allen (2009) and King (2009).)

Foucault argued that relations of violence, domination or ideology work by forcing other actors to abandon their own objectives, subordinating their own interests and strategies (Foucault 2000: 337–341). In contrast, the productivity of power involves modalities wherein one actor achieves their own objectives by structuring the field of actions through which other actors are able to pursue their own strategies. When applied to political rationalities, this idea defines relationships of 'government' as a type of joint action: government works by combining the strategic interventions of one set of actors with the strategic projects of others, through the medium of norms which structure the field of action of the latter. But this is a form of joint action that allows no room for the generative effects of communicative agreement (Boltanski and Thévenot 2006) or disagreement (Stark 2009) between the parties to this combination over shared goals and objectives. Power, in this account, is understood with reference to the varying success in bringing about one's own objectives. Honneth identifies in this concept of strategic conflict the source of the consistent elision of communicative normativity in Foucauldian analyses:

> In a social world consisting merely of situations of strategic action, something like normatively motivated consent could in no way be formed, since after all the subjects encounter one another only as opponents interested in the success of their respective aims. (Honneth 1991: 161)

Foucault's great innovation was to identify various 'distanciated' modalities through which power relationships were played out, not least by recasting the idea that the productivity of power worked through the generally distributed action of norms. The question therefore arises whether the Foucauldian approach can actually account for the possibility of strategically coordinated interaction mediated through the action of norms, without reducing the concept of 'norm' to just another term for an objectively recognized *rule* to which actors conform in every respect.

Reducing ordinarily communicative practices to the self-realizing force of socialization means that the normative dimension of interaction can only ever be located in various 'non-representational', habitual practices. As Sayer (2005: 6) observes, what gets lost in this type of understanding is 'a range of normative rationales, which matter greatly to actors, as they are implicated in their commitments, identities and ways of life. Those rationales concern what is of value, how to live, what is worth striving for and what is not'. Sayer calls these rationales 'lay normativity'; Boltanski and Thévenot (2006) and Stark (2009) refer to them as orders of 'worth'. It is this normative dimension of ordinary interaction which needs to be recovered from both the 'foundational' impulse of Habermas's search for transcendental validity conditions and the 'objectivist' impulse behind Foucault's reduction of action to the strategically governed movements of different actors.

The idea that one can only understand the relationship between individuals and wider systems of norms and regulation by taking account of 'what matters to them' (Sayer 2005: 51) is crucial to understanding the dynamics of ethical consumption. It is the only way to understand the 'responsibilization' of individuals as consumers, empowered with choice through various marketized practices of public and private provision of the means of social reproduction. If the problematization of consumption is indeed an important contemporary political rationality, then it works not through the promotion of unfettered hedonism and self-interest, but by making problematic the exercise of consumer choice in terms of ever-proliferating responsibilities and ethical imperatives.

We suggest that the notion of lay normativity can supplement Foucault's idea of ethical problematization to frame an analysis of the politics of ethical consumption. Hacking (2002: 118) has observed that 'Foucault was a remarkably able Kantian', by which he means that there is a strong emphasis in Foucault's work on the ways in which our ethical dispositions are our *constructions*. This side of Foucault, as distinct from the Nietzschean emphasis on strategic conflicts of wills, is evident in his call for a 'history of ethical problematizations based on practices of the self' (Foucault 1985: 13). The idea of ethical problematization directs analytical attention to investigating the conditions 'for individuals to recognize themselves as particular kinds of persons and to reflect upon their conduct – to problematize it – such that they may work upon and transform themselves in certain ways and towards particular goals' (Hodges 2002: 457). This concept points to the ways in which people make use of all sorts of resources to problematize their ethical selves, and not necessarily resources intended for those purposes at all. The idea of the ethical problematization refers our attention to the reasons and means through which people are encouraged and empowered to problematize their own conduct, to make a 'project' out of various aspects of their lives. This idea helps us frame our analysis of the development of policy and campaign initiatives that demand that people exercise consumer choice responsibly.

Foucault (1986: 26–28) distinguished four dimensions in elaborating a methodology to analyse modes of ethical problematization. Combinations of these processes help to constitute the aspects of individuals' lives that their moral capacities can engage with and work on. The importance of this methodology lies in placing questions of 'subjectivation' in a wider framework of object-choice, action, and rule-following that the predominant 'interpellative' understanding of governmentality has largely obscured. Foucault's four-way account of ethical problematization is a useful heuristic with which to investigate what is going on in the proliferation of discourses of responsibility and ethics around practices of everyday commodity consumption.

The first dimension of ethical problematization is the 'ethical substance' of practices of the self, 'which is the aspect or part of myself or my behaviour which is concerned with moral conduct' (Foucault 1997: 263). For Foucault, this might be one's desire, or feeling, or intentions. In the case of contemporary practices of consumption, the responsibilization of individuals involves 'choice' being defined as the 'ethical substance', that part of people's behaviour that is made the reference point for their ethical judgement.

Second, there is what Foucault called the 'mode of subjectivation': 'the way in which people are invited to recognize their moral obligations' (Foucault 1997: 264). What is at stake here is the precise *imperative* through which ethical practice is motivated: it might be accordance with divine law, or rational universalization, or simply personal conviction. In ethical consumption, the question of how obligations are recognized seems to combine different ethical motivations, but there is a very strong emphasis on implicitly consequentialist forms of reasoning governed by the avoidance of harm or the alleviation of suffering to distant others.

Third, there is 'ethical work': 'the means by which we can change ourselves in order to become ethical subjects' (ibid.: 265). This is a more practical question, and in the case of ethical consumption, the actions through which the elaboration of ethical selves is increasingly being pursued are those very humdrum, daily activities like shopping or disposing of household rubbish.

Fourth, and finally, there is the *telos* of ethical practice, the kind of self that the ethical subject aspires to become through this combination of dimensions of self-formation. In this respect, the crucial point about ethical consumption is that, both organizationally and discursively, it allows for a wide plurality of ethical positions to be pursued through a range of everyday practices loosely coded as ethical.

Foucault uses this four-way division of the dimensions of ethical problematization to suggest a rough and ready distinction between code-oriented and ethics-oriented moralities. Any 'morality', in the broadest sense, will combine these elements in different ways, but he suggested that in some, the main emphasis will be on the adherence to prescriptive codes (ibid.: 29–30). In these, the 'mode of subjectivation' is predominant, and, what is more, is given a strongly prescriptive, rule-bound inflection. In others, the focus is more on self-referential principles and actions, and it is these that Foucault (and after him, writers like Rose and Dean) dubs 'ethical'. In this sense, ethical consumption is indeed strongly 'ethical' in Foucault's terms, as distinct from being only 'moral' in a prescriptive sense.

The idea of ethical problematization helps us isolate the different dimensions of self-formation that are combined and differentiated within the broad social and cultural movement that makes up ethical consumption. Our attraction to this idea is as much methodological as conceptual. It helps to organize empirical analysis of a diffuse network of actors, actions and outcomes. In Chapter 5, we connect this idea of ethical problematization to the methodological approach developed in work on 'positioning theory' and discursive psychology (Davies and Harré 1990; Wetherell 1998). This line of work lays great emphasis on the ways in which elaborations of the self take place through various argumentative, rhetorical practices of *accountability*. It therefore provides important resources for conceptualizing how ethical consumption works as an assemblage of rationalities and technologies for the elaboration of the self, and provides a set of analytical resources for understanding empirical data on ordinary people's motivations for engaging differentially in various 'ethical' consumption practices. In short, the idea of the ethical problematization of the consumer opens up the possibility of moving beyond models of discourse as a medium through which individuals' subjectivities are 'socially constructed' from the top down towards a sense of discourse as the action through which individuals co-construct subjectivities in interaction.

In section 2.5, we argued that efforts at governing consumption do not always aim to construct the individuated consumer as the primary 'technology' for achieving their goals. We discuss this feature of ethical consumption further in the next chapter. This is not to deny that a great deal of the contemporary problematization of consumption does address individuals as consumers. But in this section, we have begun to develop an argument about *how* this individualized address to choosy consumers needs to be more carefully specified, not least by distinguishing between action, identification and subjectivity. In policy circles, and policy-related academic work, there is often an assumption that people's consumer behaviour can be regulated by providing lots of information about where commodities come from, who made them, their environmental impact, their likely health impacts, and so on. The same assumption is often at work in more politicized forms of consumer activism. Academic analysis informed by cultural theory and poststructuralism sees in these modes of address attempts to construct people's identities by having them recognize themselves as the subjects of various images of sobriety, responsibility, commitment and so on. All of these fields – policy, campaign and critical-academic – share a rather thin, atemporal conception of the self.

In fact, academic analysis might be lagging behind the most innovative practices of campaign groups, who are increasingly shifting away from the information model of consumer choice towards nuanced understandings of how people integrate their actions as consumers into broader practices of accountable self-formation. In the UK in the mid-2000s, think-tanks and campaign groups such as the Demos, the Fabian Society, Global Action Plan, the Green Alliance and the New Economics Foundation began to argue that the key to influencing consumer choice is to better understand processes of shared learning through peer groups and social networks. For example, it is argued that what is needed is a focus on the 'arts of influencing', identifying and recruiting 'intermediaries' in peer networks who persuade and influence others in conversation: 'behaviour spreads through conversations, social learning and peer group networks', and so the aim of campaigns should be to 'get people talking, inspire curiosity'.[4] We develop an analysis of this 'rationality' of governing consumption in Chapter 6. There is a significant shift underway in how influential actors seeking to shape public policy debates are themselves rethinking the most effective 'technologies' for regulating individual consumer choice. Think-tanks and policy-oriented NGOs are increasingly recommending that governing consumption in more 'ethical' or 'responsible' or 'sustainable' ways requires more than simply hectoring consumers to be more responsible and providing them with more information. What is unfolding here is an understanding that influencing behaviour can work through the classical arts of rhetoric, one which draws on technologies such as social marketing to develop programmes which focus on 'engagement' rather than 'authority'.[5] This is relevant for how we conceptualize the rationalities behind the ethical problematization of contemporary consumption, and also for how we go about empirically investigating ordinary people's engagements with these interventions in ways that do justice to their own competencies as actors and selves. In short, ethical consumption campaigning works in part by addressing *moral dilemmas* to ordinary people. In turn, the attitudes that people express about the topics raised by these campaigns need to be analysed in their rhetorical, argumentative context, rather than as expressions of pre-existing motivations or moral dispositions (see Billig 1996).

Let us return now to the theoretical terrain of advanced liberalism, governmentality and neoliberal subjectivity. The academic assumption that governmentalities work by seeking to synergize rationalities and subjectivities betrays a rather careless nominalism of the self. Generic poststructuralism of the sort that underwrites accounts of 'neoliberal hegemony' and 'advanced liberal governmentality' relies on extremely thin concepts of malleable subjectivity. The idea of ethical problematization restores some sense of how subjectivity is embedded in broader practices of self-making and personhood. This constructivist concept of personhood opens up an alternative sense of subjectivity in terms better suited to appreciating how people's sense of themselves is mediated by the routinized actions and interactions in which they are implicated. For a coherent sense of personal identity to be maintained over time, in the course of changing patterns of conduct, people must be able to integrate new events into coherent narratives. The concept of the narrative-self implies that the malleability of people's subjectivity is constrained by the degree to which new events and identifications can be integrated into ongoing storylines (McNay 2000).

The reason for splicing the idea of ethical problematization together with this discursive account of self-making lies in the inability of standard Foucauldian approaches to subject-formation to account for what Anderson (2000: 171) calls the normativity of norms. Hacking (2004: 278) argues that 'Foucault gave us ways in which to understand what is said, can be said, what is possible, what is meaningful – as well as how it lies apart from the unthinkable and indecipherable. He gave us no idea of how, in everyday life, one comes to incorporate those possibilities and impossibilities as part of oneself' (ibid.: 300). Governing through the action of norms requires an account of how norms are acted upon that goes beyond a simplistic opposition of conformity or resistance. By definition, norms do not work by being exactly conformed to. The reason for investigating the ordinary forms of reasoning that people deploy when confronted with demands to consume more 'responsibly' is that it helps better understand the 'lay normativities' through which such demands are negotiated. Norms only function as norms at all because of the capacities of people to reflect on strategies and define objectives, a self-reflexive capacity that is certainly interpretative, but above all is communicative and interactive: 'There is always an element of the discretionary, elaborative, and ad hoc about how we apply rules and schemes, for they do not define their own applications' (McCarthy 1993: 30). Governmentality theory offers us little help in understanding this discretionary aspect of the functioning of norms. Prevalent constructions of 'Foucault' encourage us to presume that this dimension of interaction is misconstrued if and when it is considered to be oriented by dynamics of communicative agreement rather than strategic interests (see Hacking 2004). Understanding the ordinarily 'ethno-methodical' communicative practices through which interaction is carried on, which allow people to interpret and apply norms, to anticipate responses and to improvise, requires us to draw on alternative theoretical resources to supplement the analytics of ethical problematization.

It is not enough to just call for more empirical analysis of how discourses are received by subjects, since without further specification such analyses easily lend themselves to an interpretation in terms of resistance and refusal. What is at stake is not just a matter of adding a bottom-up perspective to the top-down perspective of Foucauldian approaches. Our argument in this chapter has been that serious conceptual and methodological consideration of the lay normativities through which coherent narrative selves are sustained by negotiating various discursive positionings *supplements* rather than merely augments top-down perspectives on governmentality. That is, it shows that what needs to be added on to theories of governmentality – an appreciation of lay normativities and practices of situated judgement – is something missing that is really essential to the coherence of the original account itself. The overwhelmingly strategic construal of action and interaction in the analytics of governmentality cannot account for the operations it ascribes to the dynamic of strategic contestation, and this requires a reorientation of the ways in which the analysis of political rationalities and technologies of the self is formulated in the first place.

2.7   Conclusion

In this chapter, we have addressed ourselves to a set of theoretical and methodological issues arising from the use made of theories of governmentality to analyse the contemporary problematization of consumption and responsibilization of the consumer. We think that there are, indeed, a set of contemporary rationalities aimed at governing consumption in relation to particular ends, and these do help sustain the 'responsibilization' of individuals as consumers. And we have suggested that these rationalities might be usefully analysed in terms of the ethical problematization of consumption. But our recourse to these Foucauldian motifs has also led us to question some of the basic assumptions of governmentality theory. We have emphasized two issues in particular. First, governing consumption often does not aim at transforming people's *identities* or *subjectivities* in a very strong sense at all. Instead, governing consumption just as often attempts to facilitate certain types of publicly observable acts of purchase: singular *acts* that can be aggregated, measured, reported and represented in the public sphere. And second, this in turn implies that the relationship between governing actions and subject-formation requires more precise specification. In particular, we have suggested that it requires greater consideration of narrative conceptions of the self and of personhood in order to understand the recursive, reflexive relationships between routinized habitual practices and capacities to deliberate reasonably through which ethical consumption practices emerge (see also Adams and Raisborough 2008).

We have argued in this chapter that ethical consumption campaigns actively address people as agents of consumer choice, but this is not to be taken necessarily to mean they address them as self-interested, egoistical utility maximizers. They tell people that, as consumers, they are subject to a new range of moral responsibilities, but also empowered to act on these in new, innovative ways. In Chapter 4, we show that these discursive positionings of 'consumers' work by problematizing consumption, presenting people with dilemmas and conundrums. But in so doing, they presume a capacity for ordinary moral reasoning that contemporary cultural and social theory is often loath to acknowledge. In Chapter 5, we turn to the ways in which people respond to these sorts of campaigns. We find people delineating the aspects of their lives that they are willing and practically able to problematize in the myriad ways now asked of them. When people talk about what they make of these injunctions to buy fair trade coffee or organic vegetable, or boycott Nike, or recycle their beer cans, or wear coloured wristbands in support of worthy causes, they do not reveal themselves to be heroically 'active' or 'creative' consumers or perfectly virtuous citizens. Our research finds people with busy lives and torn loyalties and multiple commitments and scarce resources who do what they can, and who respond positively to initiatives to make them into more 'responsible consumers' when this can be made to fit into their own ongoing elaborations of the self. However, before turning to these issues, in Chapter 3 we develop the argument that the significance of the narrative concept of self discussed in this chapter follows from acknowledging the degree to which consumption is often much less a matter of 'choice' than is sometimes supposed by cultural and social theorists of modernity and globalization. To understand the 'becoming ethical' of contemporary consumption behaviour requires a reorientation towards practice-based understandings of the relationship between self-formation and everyday consumption.

Chapter Three

Practising Consumption

3.1 The Antinomies of Consumer Choice

In Chapter 2 we introduced the idea of ethical problematization as a way of framing our analysis of the politics of ethical consumption campaigning. We linked this concept to the idea that we need to take seriously how people create and perform particular narratives of the self, and how these narrative elaborations in turn form an important dimension of how campaigns hope to have effects. In this chapter, we relate this understanding of the narrative elaboration of the self to recent work that adopts a practice-based understanding of consumption. Practice-based theories of consumption emerge in response to an impasse in consumption studies. On the one hand, a classical sceptical conceptualization of consumerism and consumer culture, which runs from Veblen through to Bauman, positions the consumer as principally shaped by surrounding structures and contexts. On the other hand a more recent line of argument has rehabilitated the consumer as an active and creative subject. As Schor (2007) suggests, the shift to the latter perspective, shaped by ethnographic studies of consumer activity, really involves a shift in the scale of analysis, from the macro to the micro scale. The emergence of practice-based theories of consumption responds to the micro–macro framing of 'the consumer' as either a passive dupe of their own exploitation or an active hero or heroine of everyday life.

As we have already suggested, there is a widespread assumption that the primary attribute of 'the consumer' is 'choice' (see Clarke 2010). In turn, in academic, campaigning and policy discourse, it is often assumed that 'choice' is the medium through which the 'the consumer' can be persuaded to become more responsible. In this chapter, we argue that thinking about

Globalizing Responsibility: The Political Rationalities of Ethical Consumption by Clive Barnett, Paul Cloke, Nick Clarke and Alice Malpass © 2011 Clive Barnett, Paul Cloke, Nick Clarke and Alice Malpass

Figure 3.1   Responsible consumption in Bristol (courtesy of Jonathan Tooby, University of Bristol)

the politicization of consumption primarily with reference to consumer choice (whether regarded positively or negatively) might be limited in theory, and is likely to be misleading when it comes to analysing this phenomenon empirically.

As a starting point for our argument, we want to suggest that discussions of consumption, consumers and choice tend to revolve around two seemingly incompatible poles. We will illustrate this antinomy by way of two vignettes from the research project on which this book is based. Consider the photograph in Figure 3.1. One thing we did at the start of our research project was to travel around Bristol, where most of the empirical work was undertaken, taking photos of relevant places and practices. Figure 3.1 shows one of those photos. This picture captures one of the tensions within the whole field of public debate about consumption. It shows a recycling box, provided by the City Council to every household in Bristol. The people who live in this house are, obviously, very responsible consumers. It seems that after a week of heavy drinking and partying, they have made sure to put all the empty bottles and cans in the recycling box. They have even remembered to put the recycling box out on the pavement so that it is more convenient for the recycling crew to pick up. Never mind the impacts of this hedonistic excess on their own health and well-being, at least they are doing their bit for the environment.

We like this picture because it indicates some of the different imperatives to act responsibly that are now articulated through registers of consumption. It illustrates the problem of squaring the demands of equally compelling imperatives to be healthy, environmentally responsible and respectful of others, all the while leading a normal life of fun, pleasure and indulgence.

Current debates about consumption – whether around issues of sustainability, public health or sweatshop labour – suggest that people have lots of choice and discretion, and that they have to exercise this choice and discretion responsibly.

But responsibility for what?

Just oneself? Then be careful with the salt, the sugar, the chocolate, the lager.

Or responsibility for the planet? Then remember to recycle and to switch off the lights when you leave the room.

Or responsibility for distant workers? Then at least make sure that your chocolate is *Divine*.[1]

Our second vignette illustrates the other pole of the responsibilization of the consumer, and it comes from one of the various 'user group' events we attended during our research project, as part of a larger, publicly funded programme on 'Cultures of Consumption'.[2] One of these events was held in 2005 at the Food Standards Agency in London. One of the participants at this seminar worked for the Department of Health, and was working on the government's new anti-obesity agenda. He rather apologetically told the audience, made up of government professionals and academics, that he had come to this role from a job in private sector marketing. His appointment was, he said, indicative of a shift in approach by government away from assuming that all you had to do was give lots of information to people and then they would adjust their consumption habits, and towards more ambitious thinking bringing about 'step-changes' in eating behaviour. This Department of Health official was, however, a bit worried that this shift might be seen by some as an attempt to manipulate people, to get them to do things they might not otherwise choose to do by using the dark arts of marketing.

In mainstream policy discourse and public debate in the UK, the politics of consumption is framed somewhere between the dilemmas illustrated by these two vignettes. On the one hand, there is an emphasis that assumes the only way of bringing about change is by *changing attitudes* of rational consumers. But this generates confusion and potential conflict about just which attitudes should be changed. On the other hand, there is the alternative focus on *changing behaviour*. One pole seems to respect the preferences expressed in everyday practices, and trust in people's willingness and capacity to do the right thing. The other pole presumes that the main lesson to be learnt from all the evidence suggesting that the first approach does not work very well is that we need to find routes around people's preferences, opinions and desires in order to *make* them do the right thing almost despite themselves. The shared assumption of both ways of problematizing consumption is that *attitudes* are separate from *behaviour*: the challenge is either to change attitudes so that people then change their own behaviour, or to find ways of changing behaviour without engaging with people's attitudes directly at all. It is this conceptual separation which leads the Department of Health official to worry: his concern was that the chosen means of changing behaviour runs against the democratic grain of market populism, which puts a premium on respecting people's expressed preferences.

Our starting point in this chapter, then, is the thought that we would do well to stop thinking that attitudes and behaviour are separate in the first place, and that the 'gap' between them somehow has to be overcome. This is what concepts of practice teach us: they acknowledge the way in which *reasoning* and *reflexivity* are folded into the things people *do* in ordinary ways. The literature which we draw on understands a practice to be 'a bundle of activities', and any practice consists of two dimensions – activity and organization (Schatzki 2002: 70–72). In this chapter, we want to develop the case for thinking about consumption in practice-based terms. In so doing, we want to focus on the different ways in which this approach leads us to think about the place of interpretative action within practices. We argue that there is a risk that practice-based theories of engaged agency and embodied ethics simply invert the dichotomy between doing and thinking found in the attitude/behaviour division, and thereby underplay the potential of discourse-based methodologies in investigating consumption practices as a result. This chapter provides, then, a theoretical entry point for the methodological discussion and empirical analysis provided in Chapter 5.

3.2   Theorizing Consumption Practices

As part of the so-called 'practice-turn' (Schatzki *et al.* 2001), recent social theory has shown considerable interest in models of ethical and political agency that attend to affective, embodied dispositions rather than explicitly articulated, intentional meanings. This 'turn' reflects the influence of various theoretical and philosophical traditions – including ethnomethodology, phenomenology, pragmatism, Heideggerian and Wittgensteinian philosophies of mind, sociological theories of practice and structuration – all of which share a deep suspicion of any 'intellectualist' or 'mentalist' construal of human action. Theoretical perspectives that are too partial to a picture of the social world governed by rules, principles, and reasoning have become the poles against which theories of practice, non-representational theories, theories of affect and performativity, and theories of the 'more-than-human' define their own significance.

In one strand of argument within this broad school of thought, associated with 'non-representational' accounts of life (Thrift 2007), the argument is made that a practice-based perspective helps to disclose new dimensions of power and politics. Amin and Thrift (2005) have explicitly expanded on the political vision that this range of affect-sensitive, non-representational, practice-oriented theories is meant to support. They argue that this work allows us to see a series of social fields in a newly political light: first, new sources of power are disclosed by these theories; and, second, new forms of political action are revealed. One of the topics that they argue is thrown into new, political perspective in this way is the field of consumption. As they put it, 'runaway consumption in certain parts of the world clearly constitutes a problem of growing proportions, one which still has to find a convincing political language which chimes with people's everyday lives' (ibid.: 230). Having identified consumption as a political field in this way, Amin and Thrift proceed to identify the new source of political power that thinking of consumption in this way brings into view. A new 'source of power' lies, they argue, in the activities of shopping and the agency of shoppers:

> Shoppers and shopping, especially in North America, have acquired an implicit economic power which is at the centre of globalization. Without these shoppers, it is possible to argue, the Chinese economy would not be booming, together with many countries in South and South East Asia, the world financial system would be in difficulty (since consumer demand translates into demand for credit as well as all kinds of currency dealings and props), and many corporations would collapse. On the other hand, without these shoppers, we would also have a slower rate of resource depletion, less damaging global climate change, and much attenuated energy demands. It is clear that, underlying this dynamic, there is a set of ethical issues which are usually posed in much too grand a way for consumers to grasp. (Amin and Thrift 2005: 230)

This clarion call to think about consumption as a political field and to therefore think of consumers as political agents is symptomatic of a broader elision between consumption and shopping that one finds in both academic and non-academic commentary. In making the leap from the politics of consumption to the consumer as political actor, Amin and Thrift actually close down key questions about the contemporary politicization of consumption.

Amin and Thrift certainly provide a succinct summary of some of the challenges of politicizing consumption:

> One issue is how to frame environmental imperatives in ways which relate to people's everyday lives, as in recent attempts to politicise cleanliness. How is it possible to ensure that the needs of those who have very little can be met whilst curtailing the extravagances of the well-off? These issues of frugality and redistribution have to be sieved through networks of use and practice built up over many years, networks whose rightness is very often felt in the guts. A parallel issue is the attempt to produce a politics of the geography of environment and consumption, through both practical initiatives like fair trade, farmers' markets, slow food and other forms of ethical consumption and media-savvy initiatives which attempt to reconnect everyday items like food with their origins in far away places. Another area is the politics which attempts to attack consumers' love for objects such as cars, through semiotic

deconstructions which use the full force of savage irony, humour and embarrassment in order to tear the car away from consumer heartstrings. Then, there is a politics which attempts to work with the affirmative nature of many elements of consumer objects, for example, their status as gifts which trace out networks of friendship and kinship, and through these networks redefine what are acceptable exchanges, thereby hitting at the concept of value itself as it is understood in the minutiae of everyday life. (Amin and Thrift 2005: 230)

This list certainly captures the range of issues and responses involved in the contemporary politicization of consumption. What is most valuable about this account is the acknowledgement that a great deal of consumption is embedded in routines, in relationships and in deeply ingrained habits. But there is a further implication that arises from this acknowledgment which Amin and Thrift do not develop. In drawing attention to the ways in which consumption is embedded in everyday practices, it seems difficult to do away with the idea that consumption is still best thought of as a distinct practice as such, or with the sense that the primary dynamic of consumption lies in the affective attachments between consumers and objects. Thrift (2008: 21) has argued that recent research on the aesthetics of consumption practices, which locates commodity consumption in emotional networks of gift-giving and sharing, should lead us to rethink consumer capitalism as 'part of a series of overlapping affective fields'. This argument is part of a broader move to rethink commodity cultures not as worlds of fetishism but as worlds of enchantment (Bennett 2001; Foster 2005; Hetherington 2007), so that people's attachment to commodities appears not as an index of their alienation but as so many ways of engaging with the world. However, these arguments still centre our thinking about consumption on the allure and attraction felt by consumers towards discrete objects. They certainly reverse the evaluative judgement of this relationship, but they do so only by reproducing the idea that the key agent in the politics of consumption remains the consumer.

These approaches begin to recognize that lots of objects get consumed in the course of loving, befriending and relating to others. But they still conceptualize consumption primarily in terms of exchange, rather than in terms of systems of provisioning and background infrastructures that enable all these affective exchanges to go on. A large part of everyday consumption is ordinary, in the sense that it is embedded in routines which are inscribed in socio-technical systems of life (Gronow and Warde 2001). With this in mind, we want to develop an argument which, by thinking about consumption as embedded in practices, will seek to make the figure of 'the consumer' almost disappear, and then to reappear under different guises. Thinking of consumption in practice-terms should, we argue, bring into view various agents of the politicization of consumption that are not properly thought of as consumers at all, and who sometimes might even appear as reasonable, reasoning citizens. In making this argument, we want to develop the claim that the focus in cultural-theoretic accounts of consumption on the dynamics of desire underplays the extent to which a great deal of consumption remains tied to imperatives of necessity (Trentmann and Morgan 2006). If the desirous, affectively attuned subject of consumption is automatically rendered as a consumer, more or less active, then in contrast the subject of consumption-as-necessity is not best thought of as either active or passive, but as engaged in ongoing routines. It is this routinized dimension of consumption that a practice-based approach throws into relief.

What, then, does it mean to approach consumption as embedded in practices? Reckwitz (2002: 249) defines practice as 'a routinised type of behaviour which consists of several elements, interconnected to one another: forms of bodily activities, forms of mental activities, 'things' and their use, a background knowledge in the form of understanding, know-how, states of emotion and motivational knowledge'. The relevant part of this definition is the acknowledgement that material *things* play an important role in practices. This implies that we might begin to think of consumption as an aspect of all sorts of practices, not just those like 'shopping', which we all too easily recognize as 'consumption'. Material goods such as clothes, toys, food, household items like washing machines and televisions, are all wrapped up with the routines and relationships of everyday life (Miller 2009). And consumption of this sort of 'stuff' is embedded in the social relations of practices such as parenting, growing up or having a career (Gregson 2007).

The first thing that thinking about consumption in practice-terms should do, then, is to make us wary of the tendency to think that consumption is the name for a distinct set of practices, mainly to do with shopping and purchasing things, or perhaps more broadly the use of commodities for purposes of social reproduction and the symbolic or even affective mediation of identity. It is in this sense that consumption is easily offset against other practices, notably 'production'. Rather than thinking of consumption as a distinct range of practices, we follow Warde's (2005: 145) proposition that 'consumption occurs within and for the sake of practices'. Warde's approach builds on Schatzki's (1996) conceptualization of practices as activities that are composed of 'doings' (understandings of how to do things) and 'sayings' (explicit statements relating how to do something or that something is the case). Schatzki (ibid.: 98–109) distinguishes between dispersed and integrative practices. Dispersed practices are open-ended features of many activities, and include actions such as describing, walking, handwriting, listening and so on. Integrative practices are bundles of activities that make up particular domains or activities, such as cooking, motoring or being a football fan, and they contain particular combinations of dispersed practices bound together by normative ends and emotions shared amongst those performing the practice. This distinction allows us to see that consumption might not be best thought of as an integrated practice, with a subject called 'the consumer' at its centre; it might be best conceptualized as a dispersed practice: 'consumption is not itself a practice but is, rather, a moment in almost every practice' (Warde 2005: 137).

Warde argues that most integrated practices 'require and entail consumption' (ibid.), and it is in this sense that we can think of consumption as occurring 'within and for the sake of practice'. The use and using-up of material objects and resources takes place as a function of their role in coordinating and accomplishing practices:

> Appropriation occurs within practices: cars are worn out and petrol is burned in the process of motoring. Items appropriated and the manner of their deployment are governed by the conventions of the practice: touring, commuting and off-road sports are forms of motoring following different scripts for performers and functions of vehicles. (Warde 2005: 137)

This understanding of consumption as a function of practices which have their own dynamics and trajectories suggests that the conjunction of consumption and expressions of identity in so much cultural and social theory might require rethinking. Rather than assuming that consumption serves an expressive function, the idea that consumption takes place in practices implies that a great deal of it remains governed by the instrumental and purposive norms of utility, efficiency and effectiveness (Warde 2005: 147). In short, if, as Watson (2008: 6) puts it, consumer goods can be thought of as 'the "stuff" of which socialness is made', then a practice-based definition of consumption requires us to consider the relationship between the 'software' of using and exchanging goods and the socio-material 'hardware' that supports such activity.

Warde's conceptualization of consumption as a function of practices has important implications for how the politics of ethical consumption campaigning are understood. A great deal of policy and academic work continues to think of the consumer as the natural agent of consumption. For example, we have already noted above how Amin and Thrift make this leap from the politics of consumption to the consumer as a political agent. Warde's approach helps us to see that the articulation of consumption and the consumer is a contingent effect:

> People mostly consume without registering or reflecting that that is what they are doing because they are, from their point of view, actually doing things like driving, eating or playing. They only rarely understand their behaviour as 'consuming'; though, the more the notion and discourse of 'the consumer' penetrates, the more often do people speak of themselves as consuming. However, such utterances are usually references to purchasing and shopping (Warde 2005: 150)

Consumption, he suggests, 'occurs often entirely without mind' (ibid.). This begins to explain the limitations of campaigns aimed at simply telling people to consume less, consume more responsibly, or consume ethically. These information-led campaigns presume that when people *consume* stuff, *consuming* is what they are doing. At the very least, thinking of consumption as embedded in practices implies that it might be much more difficult to change people's consumption habits than is often assumed by information-led policy strategies and identity-based academic theory. Both fields tend to assume that meaning is the key determinant of consumption habits, and that people's commitments can therefore be altered relatively easily. And as we saw at the start of this section, even those theories which are not so enthralled by meaning, and which focus instead on the affective attachments of consumption, still focus on the consumer as the primary agent of such enchantments. But if 'practices are the principal steering device of consumption' because they serve as 'the primary source of desire, knowledge and judgement' (Warde 2005: 145), then this implies a different way of understanding the relevant 'change-agents' of consumption: the key questions become how people are recruited into practices, what levels of commitment they have to those practices, and how the consumption of things, stuff, and resources is embedded in these over time, in more or less path-dependent ways (ibid.).

This way of thinking about consumption as a dispersed activity distributed across so many integrated practices displaces 'the consumer' from the centre of stories about consumption. At the same time it brings into view a much broader field of consumption than just the purchase and use of consumer goods. A practice-based understanding of consumption is therefore an important step towards analysing the dynamics of ethical consumption campaigning. This understanding involves adopting a 'provisioning' perspective to the analysis of consumption, one that locates consumption in a wider 'chain of activity' that makes it possible, such as practices of production, distribution and disposal (Fine 2002). A provisioning perspective has three dimensions (Princen *et al.* 2003: 14–15). First, it involves acknowledging the social embeddedness of consumption, which implies that 'a consumer's choices are not isolated acts of rational decision making', but are related to meaning, status and identity (ibid.: 14). This first dimension of the provisioning frame restores to view what is too readily abandoned by narratives of 'individualization', where the dynamics of consumption are located in the self-actualizing capacities of individual actors:

> Embedding consumption in a larger web of social relations leads us to ask about the influences on consumption choices, including the location of power in structuring those choices. (Princen *et al.* 2003: 15)

The second dimension of this approach is to acknowledge that consumption goes on all along chains of provisioning. Consumption is now understood as resource use in networks of production, distribution, marketing, retailing and 'consumption' as ordinarily understood; it is a moment in the creation, distribution, use and disposal of material items and immaterial services (Hetherington 2004). The third dimension follows on directly from the second: consumption is not just an activity of 'the consumer' – it goes on all along these types of networks and is undertaken by all sorts of individual and collective agents. This perspective shifts attention away from individual efforts to adjust consumption behaviour, drawing into view a broader, more complex pattern of activities that generate issues which often become visible and politicized as consumption problems (Dauvergne 2008).

In this section, we have argued that if we acknowledge that 'patterns of consumption follow from and reflect the effective accomplishment of what people take to be normal routines and practices' (Van Vliet et al. 2005: 15), then this requires that we acknowledge that routines and practices are inscribed in systems and infrastructures of provision. They are not, in short, 'mere' routines or conventions, but have a dynamic that escapes the agency of individual 'consumers' irrespective of how affectively attached to particular consumer goods we suppose them to be. In short, this approach allows us to conceptually detach our understanding of consumption from the romantic vestiges of the classical sociological vocabularies of 'consumerism', which still orient accounts of the affective enchantments of capitalist commodification. Conceptualizing consumption from this provisioning perspective brings into view the ways in which the problematization of 'the consumer' in the contemporary politics of consumption is the effect of contingent strategic interventions.

3.3   Problematizing Choice

Theorizing consumption in practice-based terms should lead us to reassess the proliferation of consumer-oriented rhetoric. As Shove (2003: 15) observes, the view that 'the fate and future of the planet depends upon the cumulative consequences of what people do in their role as relatively autonomous shoppers is immensely pervasive'. However, if the figure of the consumer is a cultural-dominant of our times, nevertheless it would be a mistake to assume that the prevalence of choice-talk is just an appendage to free-market ideologies. Contemporary mobilizations of 'consumer choice' are not necessarily only about unfettered hedonistic self-interest. Rather, they are just as often about the responsibilities of self-monitoring and caring for others. The emergence of public health initiatives around childhood obesity are an example of this governmentalization of consumer choice as a vector of responsibility (see Colls and Evans 2008). Among the key agents in the splicing together of *choice* and *responsibility* are activist and campaign organizations. A practice-based approach allows us to see that the consumer is not necessarily interpellated as an agent of 'consumer choice', but is an entry point into problematizing the conventions, routines and habits which bind people into these distributed systems of provisioning (Hobson 2002). The practice-based perspective on consumption we are recommending in this chapter allows us to relocate 'consumer choice' within a distributed pattern of agency, and to reconceptualize the proliferation of choice discourse as strategically activated in the course of campaigns. In short, it allows us to recognize that the seeming ubiquity of choice discourse arises from diverse dynamics rather than simply being the successful ruse of hegemonic neoliberal ideology, and therefore opens up a space for a more discriminating analysis of the potentials and limits of these forms of mobilization.

To illustrate the different perspective on 'the consumer' which follows from adopting a practice-based perspective on consumption, consider the analysis of how choice has become a medium of political action for citizens in affluent Western societies provided by Paul Ginsborg:

> Nowhere is choice more important than in what we consume on a daily level. In everyday life, from the moment we wake up [. . .] we are making decisions about consumption: toothpaste, shampoo, conditioner, shaving cream, razor blades, toilet paper, make-up, perfume, clothes and shoes for the day, beverages, types of yoghurt and cereals, bread, spreads, jams, bicycle or car, train or bus. And all that is just to get the day started. (Ginsborg 2005: 60)

This nicely captures the sense that consumption is embedded in everyday routines of domestic and work life. But notice just where 'choice' is located here: in the decision over *which* toothpaste or shampoo to use, or which type of breakfast, or which mode of transport? These choices, if that is what we wish to focus on, are of course themselves embedded in a set of practices that Ginsborg does not foreground in this account: practices of washing, bathing, eating, being clothed, travelling, and so on. And people might have much less discretion over the meaning and material of these types of practices. There is, in short, a background context to this description of a field of ethical and political discretion, a background made up of a whole set of infrastructures of consumption, including conventions of cleanliness, practices of self-presentation, cultures of money, and the spatial form of modern urban life.

So from another perspective, there are all sorts of inconspicuous forms of consumption implicated in everyday life which might have little to do with 'choice'. The focus in policy literatures and academic theory alike on 'the consumer' – a focus which assumes that lifestyle choices, behaviour patterns and individual discretion are the key drivers of consumption patterns – can divert attention from the degree to which individuals' consumption is shaped by their practices. As Shove (2003: 4) observes, people do not really 'consume' energy or water per se; rather 'such resources are used in the process of accomplishing normal social practices and achieving a taken-for-granted standard – for example, of comfort or cleanliness'. This is one reason why focusing only on the affective attachments people might have to particular objects or commodities misses a lot of what is really going on. Even these affective attachments to particular forms of consumption are dependent on a wider set of background conditions:

> In practice, the actions and inactions of individual households are rather directly dependent upon a variety of mediating devices and upon infrastructures to which they are attached. (Shove 2003: 4)

Shove's concern is to bring into view those background practices against which Ginsborg constructs his realm of choice. Her emphasis is upon the way in which consumption is not a matter of discretion at all:

> While some actions seem optional and some figure on mental lists of things to do, many others – like showering, ironing, or washing hair – are accomplished without further thought or reflection. (Ibid.)

In turn, these conventional practices depend on a whole background of consumption:

> Contemporary routines of washing and bathing suppose the existence of taps, showers and sinks. Likewise, using electricity is impossible in the absence of things such as light bulbs, vacuum cleaners, washing machines, heaters and computers. (Ibid.)

In the first sentence of this formulation, routines are seen to depend on the existence of an infrastructure of consumption. In the second, the functioning of the infrastructure is seen to be dependent on the patterns of use in everyday life. We have here, then, a clear sense that provisioning and consumption are inextricably entwined. This means that any attempt to problematize consumption is likely to involve attempts to bring into view the ways in which acts of discretion are connected with habits, routines and wider systems of provisioning. Shove (2003: 3) holds that 'much consumption is customary, governed by collective norms and undertaken in a world of things and sociotechnical systems that have stabilizing effects on routines and habits'. In turn, consumption is 'mediated by a set of intervening devices' (2003: 9), and these devices are crucial for carrying out practices: 'patterns of consumption follow from and reflect the effective accomplishment of what people take to be normal routines and practices' (2003: 15).

The conclusion that follows from this is that analytical attention should shift from a focus on 'moments of acquisition to routines of use', and away from 'consumers' to 'practitioners'.

The practice-based perspective helps us see that a key mechanism of ethical consumption campaigning is bringing aspects of 'ordinary consumption' to mind, making background contexts explicit by rendering them problematic with reference to some set of consequences or outcomes or harms. But this effort does not necessarily require activating people as rational market-choosers. It requires finding ways of making various political *issues* – trade justice, human rights, climate change – both relevant to people's ordinary concerns and actionable in their everyday routines (see Marres 2007). Increasingly, it involves finding ways of articulating domestic spaces of everyday social reproduction with extended networks of production, distribution, use and disposal (see Bulkeley and Gregson 2009). In both Chapter 5 and Chapter 6, we show that when people reflect on their consumption habits and their roles as consumers, they do so by reference to their identifications as mums and dads and sons and daughters and brothers and sisters and friends and lovers and workmates and bosses and comrades; as Christians and socialists, councillors and counsellors, teachers and pensioners. Similarly, in terms of making these concerns actionable, various simple objects play an important role. For example, Hobson (2006a) identities an emergent 'techno-ethics' in sustainable consumption campaigns, in which objects such as light bulbs, or recycling bins activate alternative consumption habits not simply through 'shopping', but by reconfiguring the routines of domestic life. This is an example of programmatic interventions which seek to transform the infrastructures of everyday consumption.

A practice-based perspective allows us to acknowledge that consumption is a function of a whole set of infrastructures that serve as the background for more explicit forms of conduct and interaction. These might be infrastructures of feeling, like Shove's conventions of cleanliness and comfort; or infrastructures in the more clunky, 'material' sense of mediating devices like taps and pipes and wiring (e.g., Crang and Graham 2007); or the social relations of work, domestic labour and household income (e.g., Pahl 1999).

The difference that this perspective introduces is illustrated by the contrast between the account of Ginsborg, who foregrounds routine in order to identify potential activation of choice, and that of Shove, who allows us to see in consumption not just habit, but perhaps even specific forms of compulsion and obligation. Delineating the range of activities over which the potential agency of consumer choice is invoked is helpful for differentiating the forms of political strategy adopted across the range of contemporary ethical consumption campaigns. For example, the sort of choice-talk invoked by Ginsborg is characteristic of campaigns that seek to activate people's positive and negative capacity to buy or not buy certain commodities: to purchase ethically sourced products, or boycott nasty, exploitative multinationals. This sort of rhetoric and agency is one part, but just one part, of the repertoire of ethical consumption campaigning: for example, in anti-sweatshop campaigns, fair trade and organic networks, animal rights campaigns and, to an extent, in campaigns around transport and energy. However, these last two are also shading into areas where the appeal to consumer choice is much less resonant: for example, in debates about sustainable consumption in which environmental futures are at stake, the focus is less on changing the choices that people can make within existing practices, and more and more on reconfiguring component parts of whole assemblages of practice. These are the debates that Shove's type of analysis is most clearly concerned with. Likewise, campaigns around fair trade and trade justice attempt to transform whole architectures of commodity production, distribution and retailing.

The point of drawing attention to the distinction between choice-within-practices and practices themselves, represented by the arguments of Ginsborg and Shove in the preceding discussion, is not to dismiss the 'choice' repertoire as mere window-dressing or feel-good politics. The organizations worrying about sustainable infrastructures and global trade regimes are, after all, the same ones actively mobilizing people as 'consumers' to exercise choice quite deliberately in the supermarket. What the distinction draws our attention to is that the mobilization of 'the consumer' in ethical or sustainable consumption politics is not solely aimed at mobilizing the power of consumers in the market to effect aggregate changes in patterns of demand. As well as encouraging changes in purchasing behaviour, problematizing everyday spaces of consumption is also used to raise awareness, build social networks, enrol people into campaigns and publicize broad public support for organizations and campaign networks whose strategic aims remain firmly focused on bringing about changes in regulatory regimes, market structures and systems of provision (see Murray 2004).

Our argument here is that a practice-based approach to consumption is not simply something to be found in academic literatures. Far from it, there are interesting lessons to learn about practice from looking at how the politicization of consumption actually works. Rather than supposing that the politics of consumption is all about consumers' identities or even their affective attachments, or that the politics of consumption distracts from the politics of production, we suggest that there is a broader shift in governance and campaigning around consumption which increasingly seeks to transform people's everyday consumption behaviour by acting on the architecture of consumer choice. The idea that information is all that is required to ensure effective market supply in response to consumer demand for cleaner, fairer, greener products is increasingly being adjusted across a whole series of campaign fields. Instead, it is increasingly argued that key to effective change lies in providing infrastructures that support sustainable practices combined with a degree of 'self-binding' constraint arrived at through regulating 'choice-sets'.[3] Efforts to shape the choice-sets available to ordinary people can certainly be thought of as attempts to configure people's ethical conduct 'behind their backs'. But these efforts also depend on an acknowledgement that the habitual and the deliberative, the affective and the cognitive are entangled together in people's ordinary conduct: they aim to get people talking about the things they do, the feelings they bring to everyday consumption.

The argument that consumption is best approached as a kind of 'dispersed practice' often done 'without mind' helps us to begin to rethink what is going on in the emergence of a politics of consumption centred on the figure of 'the consumer', but attached to an array of issues from climate change to human rights to labour rights. This is the result of the diffuse, far from coordinated efforts of governments, NGOs, social movements and academic experts to explicitly articulate the *consumption background* of many practices, and in so doing to invite ordinary 'practitioners' to problematize their activities as consumers. In Chapters 4, 5 and 6 we expand on the argument that the discursive interventions used in ethical consumption campaigns are not primarily aimed at encouraging generic consumers to recognize themselves for the first time as 'ethical' consumers. Rather, they aim to provide information and practical devices to those people already disposed, for a wide variety of reasons, to support or sympathize with certain causes; information and devices that enable them to extend their concerns and commitments into everyday consumption practices. These acts of consumption are in turn counted, reported, surveyed and represented in the public realm by organizations who speak for the 'ethical consumer'. Ethical consumption campaigning redefines everyday consumption as a realm through which people can express a wide range of concerns and engage in a broad set of projects, including social justice, human rights, development, or environmental sustainability. The repertoires of consumerism are a means of extending existing dispositions into new areas of practice, and are related to new forms of public action by organizations concerned with a range of contentious issues.

It is certainly the case that the practice-based understanding of consumption outlined above can often still support an expert-led approach to political intervention:

> the challenge is to identify critical moments or turning points at which sociotechnical trajectories and the ways of life associated with them might be nudged, if not 'steered' in a different direction. In practice, this means looking for opportunities to modulate pathways of transition through considered forms of strategic intervention, and facilitating interaction between the many actors involved in configuring sectors, services and institutions. (Van Vliet *et al.* 2005: 18)

This formulation is still addressed to a certain sort of expert audience, giving advice on how best to intervene to change other people's behaviour. Shove (2003: 5) suggests that her practice-led, co-provisioning approach to consumption implies a shift of research and policy to 'the collective restructuring of expectation and habit'. The question that is raised by this suggestion is whether this call for 'collective restructuring' means just changing collectively held expectations and habits, or whether and how it might extend to finding ways of doing so collectively. Here there is a salient contrast with Ginsborg's rhetoric of choice, which presumes a more participatory politics of responsible choosing than one finds in much of the 'practice theory' derived from science studies. The practice-based approach to consumption, with its welcome focus on infrastructures and habits of communication, can still too easily reproduce the dichotomy between changing tastes and practices behind people's backs in contrast to providing lots of information with which we opened this chapter. Our argument in this book is that ethical consumption campaigning can be understood as a field in which one finds practical experiments that work through the difficult relationships between changing practices and extending opportunities for participation.

The ambivalence about the meaning of 'collective action' noted above in Shove's account of the relevance of a practice-based understanding of consumption for research and policy reflects a broader tendency in practice-based theories to reproduce and reverse the dichotomy between habitually reproduced mute practice and consciously deliberate action (see Barnett 2008). Here, it is useful to recall Warde's suggestion that consumption is a dispersed practice; consumption is embedded in integrated practices alongside other dispersed practices like rule-following, interpretation, judgement and so on. Rather than assuming that people have to be compelled to change their habits for their own good and the good of others, this understanding of consumption as embedded in practices implies focusing attention on the ways in which attempts are made to reconfigure routines and habits by provoking occasions when rules and conventions are bought to mind, made objects to be reflected upon, and on the practical devices and courses of action which are offered as alternatives. Judgement, interpretation and explanation – that is, rationality – breaks out when our practices and routines usages are made 'in some way problematic or in doubt' (Tully 1989: 196). Understanding practice in this sense, as combining ways of doing and ways of saying, enables us to recognize that a great deal of ethical consumption campaigning aims to do just this – to generate occasions for reflection on practices and provide resources and devices for reconfiguring habits and routines.

It is in light of this theoretical understanding that we would suggest that far from 'consumer choice' being straightforwardly championed and promoted, it is increasingly circulated as a term in policy discourse and public debate by being *problematized*. The problem of how to square the choices of free individuals with various public or collective goods is a recurrent theme of concern across various fields of policy and public debate. And across these diverse fields, the primary medium for acting on these dilemmas is not necessarily anonymous acts of consumer choice at all. Rather, judging by the rationalities and technologies deployed in policy and campaign fields, the shared focus is upon providing resources and reasons for people to act on and talk about these dilemmas in everyday contexts (see Burgess 2003).

3.4 Articulating Background

So far in this chapter we have argued for a practice-based approach to consumption, and have suggested that this follows both certain strands of academic social theory and from the observed rationalities of ethical consumption campaigning. This understanding of consumption leads us to suggest that the strategic problematization of consumption takes the form of bringing 'to mind' what are often habitual or taken-for-granted aspects of routines, in efforts to provide people with the resources with which to adopt a reflexive perspective on aspects of their everyday lives. In this section, we want to explore a little further how this practice-based approach recommends a particular methodological perspective when it comes to researching ethical consumption activity – one which takes seriously people's capacities to reason about their own practices.

The first thing we need to do is dispense with the suggestion that a practice-based perspective invalidates methodologies that make use of talk-data. Non-representational interpretations of the concept of 'practice' have too quickly become associated with the overt disavowal of 'discourse' as a conceptual or methodological resource (e.g., Whatmore 2006). Discourse has become associated with the sins of 'representational' thinking and the predominance of 'scriptural' methodologies. In this section, we want to explain why thinking about consumption as a function of practices should lead us to take much more notice of how people talk about their lives, rather than supposing that there is a better access to their practices than unreliable talk (see also Laurier 2010). One argument which suggests that thinking in terms of practices should lead us to be wary of talk-data is made by the anthropologist Maurice Bloch (1998). Bloch is concerned with rethinking concepts of culture in light of an acknowledgement of just how much of the knowledge that shapes social life is embodied and habitual. He follows the phenomenological account of embodied agency associated with Dreyfus and Dreyfus (2005), which focuses on learning and expertise as practices that involve much more than verbal articulation and interpretation of codes and rules. He takes this approach as a warrant for the broad methodological claim that 'what people say is a poor guide to what they know and think' (Bloch 1998: 3). Bloch's argument is that practical everyday knowledge is non-linguistic, and that while it can certainly be rendered into language, this changes its form so fundamentally that talk-data shouldn't actually be trusted:

> Thus, when our informants honestly say 'this is why we do such things', or 'this is what this means', or 'this is how we do such things', instead of being pleased we should be suspicious and ask what kind of *peculiar* knowledge is this which can take such explicit, linguistic form? Indeed, we should treat all explicit knowledge as problematic, as a type of knowledge probably remote from that employed in practical activities under normal circumstances. (Bloch 1998: 16)

The implication here is that talk-data is de facto suspect as a way of getting at practices. But one can arrive at this conclusion only by making two assumptions. First, by forgetting that conscious, deliberative reflection on action is part of practices, not a separate activity from them; and, second, by assuming that talk-data should always be considered as a representational medium, a window onto a subject's mind. The distance between an explicit 'version' and the practices upon which it reflects is only an insurmountable problem if one implicitly holds to an epistemological understanding of methodology which is meant to 'capture' actual events more or less accurately.

What is lacking in this sort of methodological interpretation of practice theory is any attempt to theorize the relationship between reflective accounts of practices and practices as such. There is in Bloch's account a sense that the capacity to reflect in this way is weirdly alien to practices. Why assume that it is 'peculiar' to be able to articulate reasons about actions whose logic might well be non-linguistic in form? And why should this 'peculiarity' lead to a recommendation of systematic scepticism towards informants' accounts? Only by holding fast to a norm of 'representation' for talk-data could one arrive at this 'peculiar' recommendation. Bloch recommends participant observation as the best way of getting at practical everyday knowledge, because this is patterned on the same type of non-verbal, non-rule governed 'learning' process (Bloch 1998: 25–26). The authority of epistemological representation has not disappeared here, it has just been secreted into an account that appears, on the face of it, to be resolutely non-representational at the level of its ontology of knowledge. It is not clear why observational accounts by social scientists of practice should be accorded a validity that is so presumptuously withdrawn from the reflections of actual practitioners. There is, in short, a very distinctive politics to non-representational accounts of knowledge, practice, learning and ethical expertise which recommend methods of data collection, generation and analysis that deliberately render their research subjects mute. This move is bolstered by an appeal to conceptual arguments that imply that *people don't know what they are doing*, where the meaning of *not knowing* is derived from finding the *absence* of a certain type of understanding that still remains, after all is said and done, the privileged preserve of the academic expert (see Barnett 2008).

What we have here, then, is a baby-and-the-bathwater problem. Practice-based theories of social action fold notions of intention and rationality into the ongoing flow of action (Lave 1988: 182–184), as distinct from linear models of action in which social practices are understood to be the determinate outcomes of intentions. Part of the problem in accounts which equate practice with the 'non-representational' dimensions of action derives from overestimating the degree to which embodied habitual action escapes reflexivity: 'Quite a lot of what constitutes us as who we are does not go all the way down to habits we cannot even objectify' (Eagleton 2003: 192). What is really needed is to get away from thinking of the relationships between habits and reasons in terms of spatial, depth metaphors, or temporal metaphors about priority (Barnett 2008). The issue at stake is better posed in terms of how habits and reasons are folded together.

In order to better conceptualize why the problematization of consumption works through generating reflexivity about practices, we make use of Charles Taylor's (1993, 1995) account of engaged agency and the concept of background, one which draws on themes from Heidegger and Wittgenstein to develop an avowedly anti-representational understanding of ordinary reflexivity. Background for Taylor is what 'we usually lean on without noticing' (1995: 77). In this sense, background has two conditions. First:

> It is that of which I am not simply unaware, as I am unaware of what is now happening on the other side of the moon, because it makes intelligible what I am uncontestably aware of. (Taylor 1995: 69)

Second, and at the same time:

> I am not explicitly or focally aware of it, because that status is already occupied by what it is making intelligible. (Ibid.)

The first condition implies that background is 'what I am capable of articulating, that is, what I can bring out of the condition of implicit, unsaid contextual facilitator'. This is, we suggest, a crucial aspect of any concept of practice. At first sight, this implication that background can be articulated seems to run up against the second condition, which holds that background is not an explicit, focal object for actors. But this would only follow if we assume that the whole of background, as what makes experience intelligible for us, could be articulated all at once. Any act of articulating background will itself have its own background conditions, a whole set of practices and dispositions that hold fast: 'bringing to articulation still supposes a background'. Taylor suggests that there is no reason to think that articulating background involves rationally reflecting on conditions as an abstract object of contemplation. The only way of such articulation being intelligible is by acknowledging its finite partiality:

> We can't turn the background against which we think into an object for us. The task of reason has to be conceived quite differently: as that of articulating the background, "disclosing" what it involves. This may open the way to detaching ourselves from or altering part of what has constituted it – may, indeed, make such alteration irresistible; but only through our unquestioning reliance on the rest. (Taylor 1995: 12)

There is a simple lesson to learn from Taylor's post-foundationalist account of the 'logical geographies' of habit and reflection: it is not necessary to presume that reasons that can be made explicit were present as maxims behind the actions to which they are retroactively attributed in order to be able to acknowledge that the articulation of reasons can reflexively disclose important aspects of those actions. If there is no sharp line between unarticulated know-how and explicit knowledge, then this implies that the latter should be thought of as providing a step towards acknowledging the responsibilities entailed in actions (Taylor 2000).

The strict 'non-representational' construal of the concept of practice therefore misses the force of post-foundational philosophy and social theory by clinging so tightly to a set of dichotomies between habit and reason that it claims to supersede in order to distinguish itself from 'humanist' approaches. Dreyfus and Dreyfus (2005), for example, whose account of embodied agency is paradigmatic in the field of practice theory, hold that ethical 'expertise' is enacted in the making of immediate, unreflective situational responses, and that intuitive judgement is its hallmark. It is important to emphasize where the 'rational fallacy' against which this vision of agency is presented is meant to lie. The fallacy 'consists of raising analysis and rationality into the most important mode of operation for human activity, and allowing that to dominate our view of human activity' (Flyvbjerg 2001: 23). It follows that taking practice seriously does not require doing only various forms of mute observation and disdaining talk-based research as 'representational'. Quite the contrary, practice theory opens up talk-data and key concepts such as discourse to a narrative-based understanding that helps 'develop descriptions and interpretations of the phenomenon from the perspective of participants, researchers, and others' (ibid.: 136–137).

We are now in a position to return to the field of consumption, and to consider the implications of adopting a practice-based perspective to the contemporary politicization of everyday consumption routines and habits. We argued in section 3.3 that the politicization of consumption increasingly takes the form of getting people to reflect on aspects of their consumer behaviour, and through this on their everyday consumption more broadly. To put this another way, following the discussion in this section, these initiatives encourage and enable ordinary people to *articulate the background consumption* with reference to various normative schemas – trade justice, global inequality, environmental futures or human rights. In Chapter 2, we suggested that the concept of *ethical problematization* helps to conceptualize this process; Giddens' (1984) account of the *reflexive monitoring* of conduct by selves is similarly suggestive. What both concepts point towards are the situated activities through which 'practices' are drawn into the orbit of discursive practices, are talked about, and made into fields of accountability, responsibility and decision. Problematizing consumption certainly works through trying to reattach various 'affective' dispositions such as love, compassion or sympathy to a new set of concerns. As will become apparent in Chapter 5, people are also quite capable of contesting these attributions of responsibility. And what this understanding suggests is that the 'politics' of ethical consumption might well be located at the level at which the capacities to articulate background consumption in these ways are differentially distributed, themselves dependent on background conditions of material resources and cultural capital.

In this section, we have suggested that we need to think of the process of calling commonplace practices into question and placing them in a reflective context as quite ordinary. It is a part of the habitual quality of any dispositional practice. The suggestion that a model of embodied agency embedded in routine habits invalidates a concern with discursively articulated reasons fails to acknowledge the degree to which 'routine is routinely interrupted':

> These interruptions cause people to reflect on their values and 'the proper thing to do'. This draws them into the articulation of principles. Thus, whether people *need* principles or not, they certainly engage with them. (Smart and Neale 1997: 23)

And, as we discuss in more detail in Chapter 5, this understanding of the folding together of tacit understanding and reflection in practices recommends a theoretical and methodological focus on narratives (Mattingly 1990).

3.5 Conclusion

In this chapter, we have argued that an affirmation of the value of a practice-based approach to theorizing consumption must be attentive to the pitfalls of setting up a false dichotomy between the unreflective learning of habitual expertise and fully autonomous reasoning. We have argued that a theoretical and methodological approach that thinks of consumption in practice-based terms requires us to take seriously the forms of talk through which people are invited to reflect on their own practice. The question of how much weight to give the reflections of practitioners is more than a merely academic concern, however. Striking a balance between individual freedoms of preference formation and the pursuit of individual life projects on the one hand, and justifying interventions that aim to shape tastes and bind choice in the interests of values of solidarity, autonomy, welfare and justice on the other, is a recurrent problem of democratic theory and practice. Consumption is now a central field in which this problem is played out – the consumer is at once lauded as the sovereign figure of political freedom and the most harmful agent of destructive outcomes. In turn, various examples of ethical consumption campaigning might be thought of as experimental fields in which innovative attempts to reconfigure the relationships between individual behaviour, social relations and large-scale system dynamics are being tried out.

The most interesting activist variants of the contemporary politicization of consumption are notable precisely because they move beyond paternalist models of politics as a field of expert intervention, and develop innovative forms of participatory practice. They might, in fact, be rather ahead of academic understandings of how habits and reasons fold together. The contemporary politicization of consumption is about articulating background, problematizing practices, and giving reasons all at the same time. In these fields of campaigning, one finds practical models of how to articulate the imperatives of respecting people as competent moral agents while acknowledging the web of dependent and determinative relationships into which they are woven. You also find here creative ways of mobilizing people into collective political projects which treat people not as consumers at all, even though that is what it might look like. Rather, repertoires of consumer choice serve the purpose of connecting multiple dispositions and motivations into projects which take the fairness, justice and sustainability of collective infrastructures – of resource provisioning or international trade, for example – as their target. It is for this reason that it would be a mistake to assume that the politicization of consumption implies a politics limited only to consumption. We might do well to rethink the habit of academic criticism to automatically suppose that there is a zero-sum relationship between the investment of energies in domesticated and/or lifestyle-oriented modes of civic concern and more conventional forms of collective mobilization and collective action. It is to this positive-sum relationship between various dispersed acts of everyday consumption and integrated practices of public lobbying and representation that we turn in the next chapter.

Chapter Four

Problematizing Consumption

4.1 Consumer Choice and Citizenly Acts

In Chapter 2, we introduced the idea of ethical problematization to understand how and why ethical consumption campaigning is addressed to 'the consumer'. In Chapter 3 we developed the understanding that thinking of ethical consumption campaigning in terms of ethical problematization discloses the extent to which consumption is a functional attribute of practices. In this chapter, we further develop these arguments in a more empirical register, by examining *who* is doing the problematization and *how* this process works by circulating various 'moral dilemmas' through the public sphere. We have already argued that much of the academic literature and policy discourse on the politics of consumption starts from the question of how to motivate consumers to change individual or household consumption behaviour. In Chapter 2 we suggested that this way of problematizing *consumption* by reference to the attitudes and behaviour of *consumers* takes for granted a relationship which needs to be subjected to critical analysis. In this chapter, we look at the ways in which ethical consumption campaigns work by re-articulating one of the most powerful storylines of contemporary public discourse, that of 'globalization'. We show how in ethical consumption campaigns, globalization is presented as simultaneously providing people with opportunities for innovative engagements as consumers just as it implicates them in an ever-expanding range of consequential entanglements.

Academic analysis of consumer identities mirrors broader policy discourses that focus on information, awareness and individual consumer choice. This focus tends to overplay the extent to which people's affective investments in consumption *practices* are malleable. It also tends to obscure the extent to which a great deal of consumption has little to do with

Globalizing Responsibility: The Political Rationalities of Ethical Consumption by Clive Barnett, Paul Cloke, Nick Clarke and Alice Malpass © 2011 Clive Barnett, Paul Cloke, Nick Clarke and Alice Malpass

consumer choice but is, rather, determined by the organization of collective infrastructures of provisioning. For these reasons, the predominant emphasis on information-led strategies aimed at changing consumer behaviour is increasingly questioned in theory and practice (e.g., Eden *et al.* 2007; Global Action Plan 2004; Hobson 2003; Slocum 2004). Building on practice-based understandings of the politics of consumption, this chapter explores the contingent articulation of discourses of consumption with discourses of consumerism. We investigate the discursive repertoires deployed in ethical consumption campaigning in the United Kingdom over the past two decades. We identify the distinctive political rationality of this sort of campaigning. In so doing, we challenge the a priori assumption that ethical consumption substitutes a privatized and individualistic form of action that is at odds with public and collective modes of participation.

Since the 1990s, the shared terrain of policy debates in the UK about consumerism, citizenship and the public realm is the assumption that consumerism represents a culture of individualized, egoistical self-interest. For example, under 'New Labour' governments in the late 1990s and early 2000s, proponents of market-led reforms of public sector institutions held that extending 'choice' was the only way to secure the long-term legitimacy of public services in a context where people's identities and loyalties are no longer defined by reference to work and the labour market but by what they buy. At the same time, commentators and academics argued that public life itself was in decline as people shrunk away from public participation and civic engagement into more privatized, consumer-led lives (Marquand 2004; Lawson 2009). The result, it was argued, is the eclipse of collective dimensions of citizenship and the conflation of the collective determination of shared public interest with the market-mediated aggregation of private preferences.

The supposedly depoliticizing effects of consumerism are thought to be particularly problematic because, just at the moment when 'the consumer' seems to have triumphed as the epitome of modern living, so consumption itself has become an increasingly problematic realm of contemporary governance. In campaigns around climate change and environmental sustainability, public health, and global poverty, excessive levels of material consumption in the West are identified as causes of various harms: environmental degradation, personal illness and socio-economic inequality. From the perspective that sees the rise of consumerism as a fundamentally depoliticizing trend, policy approaches and public campaigns that address people as consumers only compound the real problem – what is really needed is a reinvigoration of a more collective, republican form of citizenship.

Our starting point in this chapter is the feeling that current academic and public debates which set the egoistical, individualized 'consumer' against the virtue of the collectively oriented 'citizen' might well overlook how new 'acts of citizenship' (Isin and Nielsen 2008) are currently being configured through creative redeployment of the repertoires of consumerism by various agents of 'non-governmental politics' (Feher 2007). At first sight, ethical consumption seems to fall under the description of what Pattie *et al.* (2003a, 2003b) call *individualistic activism*, as distinct from both *contact activism* and *collective activism* (see also Li and Marsh 2008). It involves relatively anonymous individual acts, as distinct from acts which aim to contact people in authority or those which involve participating alongside other people. But there is no need to see different modes of civic engagement as mutually exclusive. People who engage in individualistic activism such as ethical consumption 'are no more or less likely to engage in collective activities or to contact the authorities than those who are not "individualistic activists"' (ibid.: 448). In the analysis that follows, we demonstrate that ethical consumption actually combines elements of all three of these ideal types of civic engagement.

As we argued in Chapter 2, our analysis of ethical consumption in the UK investigates both the *repertoires* and *agents* of participation involved in this movement. In this chapter, we flesh out both dimensions through a consideration of the campaigning repertoires characteristic of ethical consumption organizations in the UK over the past two decades. We focus in particular on the way in which these organizations are effectively 'globalizing the consumer' by providing practical and narrative pathways to people to act as 'ethical' consumers. As Goodman (2004) has illustrated in the case of fair trade campaigns, consumers are consistently urged to position themselves globally, both as affluent members of the global North and as caring and responsible towards those living in the global South. In focusing on the processes through which people are mobilized as subjects of various 'global' responsibilities, we argue that the process of *mobilization* has a double resonance: it involves engaging with the practices of people who are already supporters or sympathizers with certain causes; but it also seeks to represent their expressed preferences as 'ethical' subjects to other actors involved in making markets including state agents, corporations and regulatory agencies. In section 4.2, we flesh out the genealogical conception of 'the consumer' introduced in Chapter 2, one which emphasizes how the coincidence of the contemporary problematization of consumption and the proliferation of discourses of 'the consumer' comes about through the strategic efforts of various actors to engage people as 'consumers'. In sections 4.3 and 4.4, we identify the repertoires of engagement and participation that are characteristic of the emergence of ethical consumption, ones which lay heavy emphasis on information, knowledge and narrative storylines as ways of enrolling people into various forms of global responsibility.

4.2   Articulating Consumption and the Consumer

Critiques of consumerism tend to obscure what is most distinctive about the ways in which discourses of choice currently circulate in policy and public debates. They accept at face value that 'consumer choice' is simply a matter of egoistical self-interest promoted by rampant neoliberalism.

However, as we argued in Chapter 2, there is an internal relationship between discourses of *individual choice* and discourses of *individual responsibility*. Rose (1999) argues that the prevalence of consumerism as a political rationality has its roots in the 'de-socialization' of modes of governing, whereby it becomes possible to govern people by regulating the choices made by autonomous actors in the context of their everyday, ordinary commitments to friends, family and community. In this way, consumption becomes a new vector for governing society 'through the "responsibilized" and "educated" anxieties and aspirations of individuals and their families' (ibid.: 88). On this understanding, consumption is transformed into a medium for making up ethical selves, not in the sense of conforming to externally imposed codes of conduct in the name of collective good, but in the sense of 'the active and practical shaping by individuals of the daily practices of their own lives in the name of their own pleasures, contentments and fulfilments' (ibid.: 178–179). From this perspective, informed by theories of governmentality, discourses and practices of consumerism are central to the programme of responsibilization.

We emphasized in Chapter 2 that the articulation of 'choice' and 'responsibility' around the figure of the consumer needs to be understood as a contested process that involves the efforts of a diverse set of actors pursuing plural ends. It is not reducible to a single dominant logic of 'advanced liberalism' any more than it is a functional requirement of 'roll-out' neoliberalization. Responsibilization is evident in a number of fields, in which the relationship between states, markets and individuals is reordered:

> So whereas in the domain of health a discourse of the 'unhealthy Western' lifestyle has moved towards an individualized monitoring of health risks (with all the practices that come with it, such as fitness, healthy food and self-monitoring), the environmental sphere sees the emergence of individualization of food risks through the introduction of labelling and web-based information services. (Hajer and Versteeg 2005: 180)

The proliferation of 'consumer choice' is indicative of the modularization of a new rationality of governing through individualization. The exercise of choice becomes a basic element of 'the subjective meaning of consumption for the ordinary individual in their everyday life' (Miller and Rose 1997: 18). According to Rose, in this move the very nature of individuality is transformed along the lines of consumer choice, so that individuals are thought of as 'not merely "free to choose", but *obliged to be free*, to understand and enact their lives in terms of choice' (Rose 1999: 87). Individuals are, it is argued, reconfigured by being offered an identity as 'consumers':

> In the name of themselves as consumers with rights they take up a different relation with experts, and set up their own forms of 'counter-expertise', not

only in relation to food and drink and other 'consumables', but also in relation to the domains that were pre-eminently 'social' – health, education, housing, insurance and the like. (Ibid.)

Experts – advertisers, market researchers, psy-experts of various sorts – become crucial to this new regime of conduct, acting as 'concerned professionals seeking to allay the problems, anxieties and uncertainties engendered by the seemingly so perplexing conditions of our present. They operate a regime of the self where competent personhood is thought to depend upon the continual exercise of freedom, and where one is encouraged to understand one's life, actually or potentially, not in terms of fate or social status, but in terms of one's success or failure acquiring the skills and making the choices to actualise oneself' (ibid.). In the terms of governmentality theory, then, being a consumer is now a vector for the ethical formation of selves.

As we suggested in Chapter 2, we are broadly sympathetic to this diagnosis of the contemporary politicization of consumption, with two provisos. First, the responsibilization of consumption needs to be understood as more than a top-down process. Second, the processes of subject-formation implied by this analysis need to be understood as much less determinative than is often assumed in governmentality theory. In this chapter, we look at how efforts at governing consumption engage creatively with people's existing ethical dispositions. This focus on dispositions rather than subjectivities follows from the empirical observation that far from 'choice' being straightforwardly championed and promoted, it is increasingly circulated as a term in policy discourse and public debate by being problematized. In short, the problem of how to ensure that the choices of putatively free individuals are exercised responsibly – in terms both of those individuals' own good and the good of broader communities – has become a recurrent theme of contemporary public debate. For example, choice is problematized in terms of the potential of increased individual choice to conflict with public interest goals of sustainability and conservation; in terms of increased choice leading to greater anxiety and reduced quality of life, even reduced levels of happiness; in terms of the limitations of choice to increase or maintain equity in social provision and access to public services. In short, choice circulates as a term of public debate only in and through this register of responsibility for the self and for others. The standard interpretation of 'neoliberalism' misses this by persisting in thinking of individualism as necessarily equivalent to egoism (e.g., Harvey 2003). What is most distinctive about the contemporary discourse of consumer choice is that it focuses less on questions of choice as a vehicle for efficient allocation than it does on concerns with legitimacy, trust and capacity building (Barnett 2009).

In the UK, the problematization of choice has become most evident in debates about smoking, obesity and other health related issues in which the extension of choice in consumer markets is seen to lead to deleterious effects not just on individuals but also on the fabric of collective life itself. It is in this context that a series of discourses of 'soft paternalism', exemplified by the popularized behavioural economics of Thaler and Sunstein's (2007) best-selling *Nudge*, have emerged into public debates. In this set of debates, the concern is with how to ensure that the exercise of choice does not have a negative impact on the consuming self by regulating such choice through 'liberal paternalist' measures. Our focus in this chapter is with a distinct although related set of debates in which issues of choice are related to a set of more other-regarding concerns, with environment sustainability, global warming and trade justice. As Sassatelli (2006: 236) argues, the movements involved in what she calls 'critical consumerism' frame choice 'less in terms of rights and more in terms of duties'. In this chapter, we look at the ways in which this framing of consumer choice is related to a campaigning rhetoric in which globalization is presented as empowering people to act on certain global responsibilities through the medium of their consumption activity. We critically assess the discursive field populated by a set of think-tanks, consumer organisations and campaign groups. These include The Future Foundation (a commercial think-tank dedicated to understanding the future of consumerism); the New Economics Foundation (a sustainable economy think-tank); the Co-operative Bank (which has its own distinctive ethical stance on social responsibility and ecological sustainability); the National Consumer Council (a lobbying group for all consumers); the Green Alliance (a think-tank on sustainable development); and the Fabian Society (a political think-tank). All of these organizations regularly engage in public debates about consumption, sustainability, environmentalism and social responsibility. We discern in their activities a distinctive mode of problematizing choice as a means of recasting the responsibilities of consumers in collective and shared ways rather than in individualizing ways. Our analysis suggests that in so far as the normative discourse of markets and consumerism is rhetorically associated with paternalist discourses of responsibility, then this problematization of choice involves a double movement in which the individualization of responsibility opens up new possibilities for collective action through the medium of markets and the repertoires of consumerism.

We have already argued that theories of governmentality throw light upon important aspects of contemporary consumption practices, but that they under-theorize issues of agency. In particular, these theories tend to assume that the subject-effects implied or aimed at by programmes of rule actually come off in practice (see Pykett *et al.* 2010). There is something a little too neat about the shift in modes of governing that this approach identifies. For all the emphasis on 'contingent lash-ups of thought and action', there is a strong sense that projects aimed at governing conduct actually work. This observation certainly implies the need for more 'dialogic' approaches to the relationships between programmes of rule and practices of subject-formation. But more than this, it requires a reconsideration of whether these sorts of programmes do, in fact, aim at interpellative subject-effects at all – whether they seek to make subjects strongly identify with particular images of themselves through mechanisms of recognition. It is this latter, more fundamental theoretical and methodological question which we pursue in this chapter.

So far, we have argued that if consumerism is indeed an important political rationality in the contemporary conjuncture, then it works not through the promotion of unfettered hedonism and self-interest, but by making problematic the exercise of consumer choice with reference to various, ever-proliferating responsibilities. In Chapter 2, we discussed the idea of ethical problematization as a way of analysing this phenomenon. We argued that people are increasingly expected to treat their consumption practices as subject to all sorts of moral injunctions: they are expected to do so through their capacity to exercise discretion through choice; in the everyday activities of social reproduction mediated through commodity consumption; and in relation to a very wide range of substantive concepts of the good life. An example of this type of problematization is provided by the claims made by the Ethical Consumption Research Association (ECRA), which publishes the *Ethical Consumer* magazine. ECRA understands its audience as political actors who use their daily purchasing as votes to register their approval for certain objectives and to help make corporations accountable. In this understanding, consumer choice is presented as a medium of 'democratised morality',[1] in the sense that people now have choice about their own moral conduct and principles, and with this comes the 'need to make their own decisions, rather than follow established norms'. Here, then, we can see the process of ethical problematization of consumer choice made explicit. This campaign organization presents choice not just as a medium for the expression of moral preferences, but also as the very mechanism through which people constitute themselves as moral agents in the first place.

In the rest of this chapter we focus on two aspects of the problematization of consumption and consumer choice. First, in section 4.3 we examine policy documents on public service provision, think-tank reports on sustainable consumption, consumer reports and research polls on ethical consumers, and campaign materials of ethical consumerism organizations. Our analysis identifies the distinctive discursive registers in which consumers are addressed as bearing *responsibility* both for their own choices and the effects of their choices on others. Second, we argue that this is not simply a register of moral exhortation. Rather it reflects an explicit concern with rethinking the 'the art of influencing'[2] consumer behaviour by deploying various practical devices and strategies. In section 4.4 we show how these include education campaigns, learning about and utilizing network hubs, labelling and certification campaigns, and various ways of linking consumer purchases to opportunities to engage in campaigns. What can be discerned in this field is an emergent rationality that holds that the best way of influencing people's behaviour is to deploy the classical arts of persuasion. This finding is relevant for both how we conceptualize the rationalities behind the ethical problematization of contemporary consumption (the task of this chapter), and also for how we might go about empirically investigating ordinary people's engagements with these interventions in ways that do justice to their own competencies as persons, and not just subjects (the focus of Chapter 5).

4.3   Mobilizing the Ethical Consumer

We use the concept of mobilization here in two analytically distinct senses which are strategically related in campaigning practice. First, in this section, mobilization refers to the ways in which organizations enrol existing supporters and sympathizers into new modes of campaigning. In the next section, mobilization refers to how myriad discrete acts of purchasing are represented in the public realm as indicative of coherent trends in consumer preference for more 'ethical', 'responsible' forms of production, distribution and provisioning. The shared sense of mobilization in both cases is meant to emphasize the contingent 'lashing together' of the consumer and the politicization of consumption.

A good starting point for an analysis of the rationalities of ethical consumption campaigning in the UK is to investigate the proliferation of 'How to' guides published in this sector and the numerous 'What to and what not to buy' publications. They include books such as the New Internationalist's *Do the Right Things!* and the *Rough Guide to Ethical Shopping*[3] as well as regular magazines such as *The New Consumer* and *The Ecologist*, which also publishes *Go Mad! 365 Daily Ways to Save the Planet*.[4] These guides conform to a broader rationality that holds that the key to altering consumption patterns lies in providing information to individual consumers so that they can then change their own behaviour through exercising 'responsible' consumer choice. Accordingly, they are also routinely linked to relevant web sites where readers can find out more about various products, issues and campaigns. This focus on information provision appears to confirm the idea that ethical consumption is a reconfiguration of standard models of consumer sovereignty and market choice which are understood to be constrained only by lack of information. On this view, the role of pressure groups and campaign organizations is to reconfigure market relations by providing wider and different sorts of information to consumers. This is illustrated by one of the first guides to ethical consumption in the UK, originally published in the late 1980s, *The Green Consumer Guide*. Its starting premise is succinct and to the point:

> Clearly, if the relevant information is presented in the right way, then more and more of us will become sufficiently interested to take action through our day-to-day decisions.[5]

The same kind of assumption appears to underlie the work of one of the leading ethical consumption organizations in the UK, the Ethical Consumer Research Association (ECRA), publisher of the *Ethical Consumer* magazine. ECRA's mission statement explicitly registers an organizational commitment to the transformative role of information provided to individual consumers:

> Most consumers feel they are both in a position to influence corporate behaviour, and desire to do so, but lack the facts necessary to make informed purchasing decisions.[6]

However, before we simply assert that these efforts are part of the same agenda of individualized choice characteristic of many contemporary policy initiatives, we should note that none of these publications have mass-market sales. They circulate among distinct niche markets, targeted at people who are already likely to support certain campaigns or sympathize with certain causes. Sales are based either on personal or institutional subscription or purchases from specialist retailers, such as alternative bookshops or organic food retailers. Participants in two of the focus groups convened as part of our research on ethical consumption in and around Bristol were recruited through an advert placed in *Ethical Consumer*, and these informants expressed a clear sense that this sort of publication was something they *used* in particular sorts of ways as part of their existing commitments:

TIM: Yeah I've got a subscription to it as well. I would say it's not my main source of information about campaigns, but it is about consumption, if you're going to buy something, buy a stereo or something like that.

SANDRA: Yeah.

TIM: Then, yeah, I'll go to their guide and see what the best choice to buy is, definitely. Um, I dunno, I don't even, to be honest, I don't even see it as a magazine, because it's not much of a fun read because . . . [Laughs] . . . it's really depressing; it's about very depressing companies doing very depressing things so, I see it more as a functional tool in my life rather than as a magazine that I read.

SANDRA: Well, I read it, you know, from beginning to end. [Laughs] All of the little . . .

TIM: I like the cartoons, they're the only bit I like, enjoy [Laughs], . . . but the rest of it . . ., yeah.

SANDRA: And the letters.

TIM: It's just functional. It's just like, right, I'm going to buy something, I'll go to *Ethical Consumer* and I see what are the choices.

SANDRA: Yes, yeah.

TIM: . . . and then that's how I use it.[7]

The audience for these publications is, in short, self-selecting. These sorts of publications are aimed at empowering sympathetic readers to act on the basis of their ethical and political dispositions (see Berry and McEachern 2005). They function as a means of maintaining and extending the mobilization of people already geared to taking certain dimensions of their everyday consumption as an object of explicit reflection, as well as providing them with informational and narrative resources to help them recruit new supporters from within their own social networks.

ECRA is the most significant campaigning organization in this field in the UK. It provides information about which products count as ethical and where consumers might find them. Its activities include an extensive web site containing research reports on different sectors, companies and products; a bi-monthly magazine, *Ethical Consumer*, first published in 1989; and the co-published *The Good Shopping Guide*, first published in 2002. *Ethical Consumer* magazine includes feature analysis of specific types of products produced by different companies, ranging from carpets, clothing, computers, televisions, banks, pensions and investments, all the way to shampoo and butter.

The Good Shopping Guide condenses surveys and ratings data into an easily digestible book-size package. The rationale behind this mainstream publishing venture is clearly stated:

> By using this book you will discover more than you ever knew about what goes into the goods you buy or are thinking of buying. You will have the information you need to make clear decisions, either to buy the products of progressive and green companies or to boycott those of unethical companies.[8]

ECRA's own understanding of the scope of this sort of publication is quite modest. Rather than assuming that readers do, in fact, shop with these 'scores' in mind, the objective has been to raise awareness of issues and to get people talking, both in everyday life and in more formal public arenas.

These publications seek to provide information to existing supporters and sympathizers, as a means of empowering them to 'choose' differently and more generally to raise awareness and generate debate. The focus on providing information for these purposes is associated with specific understandings of what 'being ethical' actually involves:

> Ethical consumption, put simply, involves buying things that are made ethically by companies that act ethically. Ethical can be a subjective term both for companies and consumers, but in its truest sense means without harm to or exploitation of humans, animals and the environment.[9]

In practice, the evaluations undertaken by ECRA encompass a diverse set of ethical and political convictions. Companies are rated according to criteria which include standards of environmental reporting, pollution records, use of animal testing, recognition of workers' rights, support for oppressive regimes, irresponsible marketing and donations to political parties. This form of action is not solely focused on changing individual consumer behaviour. It is, rather, indicative of a type of 'politics of shame' in which one set of collective actors (campaigns, NGOs, charities) engage with other collective actors (retailers, suppliers, corporations) through the real and discursive figure of 'the ethical consumer'.

Ethical consumption campaigning seeks to connect the forms of care and concern already embedded in everyday consumption practices into wider networks of collective solidarity. This involves a combination of innovative devices such as the shopping guides and purchasing indices outlined above, but also the generation of narrative frames in which mundane activities like shopping can be re-inscribed as forms of public-minded, citizenly engagement. A distinctive feature of ethical consumption campaigning is the degree to which the solidarities and concerns mobilized in such campaigns are relatively de-territorialized when compared with previous examples of consumer activism. Ethical consumption seeks to connect the activities of everyday, domestic social reproduction – shopping, doing the laundry, preparing dinner – to a range of 'big' public issues such as human rights abuses, labour rights, environmental sustainability or global trade justice. This blurring of the public/private distinction is in turn, we suggest, related to the 'transnationalization' of responsibilities addressed to the potential subjects of ethical consumption campaigns. Furthermore, the targets of claims-making by ethical consumption campaigns are not restricted to national governments. More often than not, they directly address business corporations or international regulatory institutions. In part, this helps to account for the appellation 'ethical' in the UK, in so far as the motivations and justifications that circulate through these practices tend to be based less on a political vocabulary of reciprocal rights and obligations and rather more on a vocabulary of responsibility, compassion and care. What interests us here is the fact that this explicitly ethical register is articulated as part of a narrative of the declining significance of national governments and national politics more broadly.

One criterion for the analysis of ethical consumption campaigning is whether this mobilization of the figure of the empowered consumer is made in ways that connect with forms of collective, participatory engagement, or whether it wittingly or unwittingly reproduces a marketized discourse of privatized, anonymous choices. This is an important tension within the broad movement of consumerized activism of which ethical consumption is a part (see Littler 2005).

It is certainly the case that a great deal of ethical consumption campaigning takes the rhetoric of the 'hollowing-out' of the nation-state and turns it into an empowering address to consumers to realize their new-found influence. In her guide to responsible action in a globalized world, Anita Roddick, founder of The Body Shop, suggested that 'the most powerful bodies in the world, the World Trade Organization, the World Bank and the International Monetary Fund, are also the least democratic and inclusive. The result has been a major democratic deficit'.[10] Having attributed both power and opacity to such international bodies, she turns her attention to multinational corporations and national governments: 'Business itself is now the most powerful force for change in the world today, richer and faster by far than most governments'.[11] There is a two-step move involved here. First, it is asserted that all effective power is now concentrated in the hands of global economic actors, whether businesses or international regulatory institutions. Second, the response to the implied crisis in accountability that follows from this redistribution is already at hand, generated by the very same forces that give rise to the initial problem – people are now empowered *as consumers*.

The argument that power has moved from accountable national governments to unaccountable international bodies and multinational corporations is a recurrent trope in the 'What to and what not to buy' guides we described in the previous section:

> One of the implications of free trade and globalization is that we have seen an increase in the power of multinational corporations who, with their huge capital resources, have become nomadic. They are able to move from country to country seeking out new, more profitable opportunities. Moreover, they are often subsidized by governments keen to encourage capital flow into their countries. Whilst we are witnessing the globalization of business, we see little evidence of the globalization of government able to keep control of abuses of economic power. Indeed, there has been a reduction in the role of government in the areas of both economic and social policy, with increased emphasis on the free market as the main mechanism for development.[12]

This account characterizes globalization by juxtaposing capital mobility and reactive governments:

> Large corporations have a significant advantage over governments. They are able to cross borders much more easily. The transnational corporations with their massive stocks of private capital are much more influential on the global stage than any government or even intergovernmental agency can be.[13]

This type of rhetoric reproduces rather than contests a well-established discourse of globalization as a clear-cut shift from state-regulation to market-regulation of economic affairs. In these examples, the empowerment of 'the consumer' as an ethical actor is placed firmly within what one might call a 'neoliberal' frame which takes for granted the natural operations of markets, price signalling and the aggregation of preferences. These sorts of accounts seem, then, to confirm Littler's (2005) argument concerning the lack of reflexivity in much of what she characterizes as the 'anti-consumerism' movement.

However, other organizations adopt a stance in which the individual and collective dimensions of 'ethical' action are not seen as substitutes for one another but are aligned as part of a broad movement of mobilization in markets, public spheres and formal political arenas. And this involves an alternative interpretation of the discourse of 'globalization' and its deployment in ethical consumption campaigning. In more activist forms of ethical consumption campaigning, invoking the disjuncture between the global scales of corporations and markets and the national scale of formal political participation is not deployed simply to lament the decline of the nation-state, to bemoan corporate domination, or to celebrate the power of the individualized consumer. This disjuncture is deployed, rather, to bring into view the newly empowered consumer-activist, now able to leverage their purchasing power against corporations potentially vulnerable to 'No-Logo' styles of political campaigning (Klein 2000). People are addressed in this genre as consumers *and* citizens – as citizens of the world by virtue of their status as consumers.

This framing of this globalization as empowering people as consumers is quite explicit in the activities of ECRA. They attribute the rise of ethical consumption quite directly to globalization and the deregulation of markets by national governments which have led to the increasing dominance of 'unelected' multinational corporations:

> Globalization means that people concerned about social or environmental issues can no longer, in many cases, just lobby their own government for regulatory solutions. The UK government simply has no power to ban child labour in Pakistan or to halt logging in Amazon reserves.[14]

But because of this state of affairs, so the argument goes, campaign groups have increasingly looked for active consumers to put pressure directly on companies and corporations:

> Globalization has brought about a huge increase in product choice which has significantly increased the power of consumers in modern markets.[15]

And in turn, this is used to explain the growth of consumer-oriented campaigning across a diverse range of causes and issues:

> We are fast approaching the situation now where it is unusual to find a pressure group without some kind of 'consumer awareness' campaign aimed at influencing corporate behaviour.[16]

Here, then, globalization is presented as empowering consumers not in a narrowly individualized sense, but by providing a sense of distributed activity towards shared goals.

The metaphor which is most frequently used to describe this new form of consumer power is that of 'voting':

> We don't have to feel powerless about the world's problems. Our till receipts are like voting slips – they can easily be used constructively [. . .]. If you care at all, it's really simple to do something about these difficult issues, just by making good choices while you're out shopping.[17]

It seems here that the classically 'political' function of voting has transmuted into an essentially 'economic' function of exercising consumer choice (see Dickinson and Carsky 2005). This is precisely the sort of move that is bemoaned by critics of contemporary consumerization of politics and the rise of the citizen-consumer. But this would be to ignore how this rhetoric actually aligns voting and shopping. The invocation of voting is the point at which the wider articulations of this sort of consumer-oriented activism become visible. Ethical consumption campaigning in the UK tends not, in fact, to present consumer activism as a substitute for other forms of political participation. Consumer activism is presented as augmenting the repertoire of actions already available to ordinary people for engaging with the wider world of power and influence. *The Green Consumer Guide* made this clear when it was first published over two decades ago:

> Don't forget how important it is to let other people know about the issues. Write to your local newspapers and to the national press. Contact your MP. And if local issues are your target, get in touch with your local councillors and with the relevant local government department, water authority or central government. Above all, join relevant campaigning or lobbying organizations.[18]

Here ethical consumption is framed as just one part of a broad repertoire of actions, a repertoire that combines elements of what Pattie *et al.* (2003a, 2003b) refer to as individual, contact and collective activism. The sorts of 'global feeling' that are mobilized by ethical consumption campaigns aim to sustain collective participation in networks of national and local politics:

> Ethical buying is not a substitute for other forms of political action. Nor is it necessarily just concerned with individual consumers. 'Ethical purchasing' is, for example, already being organised by clubs, societies, campaign groups, trade unions, private companies, local authorities and national governments.[19]

For ECRA, as well as other organizations in the field such as Traidcraft, Friends of the Earth or Labour Behind the Label, ethical consumption is as much about mobilizing churches, schools, trade unions and other collective associations as it is about addressing individuals as members of privatized households. Campaigns are designed to reach people as members of these sorts of associations and they encourage people not only to shop but also to join, socialize and organize. We discuss this process in more detail in Chapter 6. What we want to underscore here is the way in which the identity of 'consumer' is discursively mobilized by organizations in order to make visible to people various *acts* which they might be motivated to undertake because of much 'thicker' forms of identification: as good Christians, as concerned parents, as trade unionists, as professionals, as members of solidarity networks, as environmentalists, or as residents of particular places.

In this section, we have seen that ethical consumption campaigns and organizations re-inscribe the discourse of 'globalization' into an affirmative narrative in which people are empowered in new ways through their role as consumers. We have emphasized the narrative qualities of these campaigns to underscore the degree to which the rationalities of these campaigns aim to provide new resources for the discursive elaboration of self-identity and social practices. There are two dimensions involved in this re-inscription of globalization. First, it involves a claim that people are now implicated in much more extensive spatial networks of exchange, exploitation and advantage, so that this narrative ascribes to people a much broader range of responsibilities: to the environment, to workers in distant sweatshops and so on. Second, this ascription of responsibility by virtue of individuals' implication in the global market turns out also to provide the medium through which people are told that they are empowered to act on these new responsibilities: as consumer activists, or perhaps shareholder activists. The organizations involved in ethical consumption simultaneously make it possible for people to recognize themselves as *consuming subjects* and as *responsible subjects*; that is, to recognize themselves as bearing wide-ranging, spatially extensive responsibilities *and* having the potential for action-in-concert with others by virtue of their capacity to exercise discretion over whether or not to buy and invest in particular goods and services.

4.4 Articulating the Ethical Consumer

This section discusses the second sense of mobilization of the consumer in ethical consumption campaigning noted above, where mobilization refers to the ways in which myriad discrete acts of purchasing are represented in the public realm in order to make claims about trends in consumer preference for more 'ethical', 'responsible' forms of production, distribution and retailing. In the previous section, we looked at the ways in which the provision of information is used as a means of engaging people as subjects of self-consciously responsible choices, then this effort is intimately connected to the efforts of the same organizations to generate information about consumers as a way of representing 'the consumer' in wider public debates.

Concerted efforts at simultaneously *making the ethical consumer visible* and *speaking for the ethical consumer* in the public realm involve the deployment of various calculative technologies of the sort we discussed in Chapter 2. Governmentality theory draws attention to the ways in which governing dispersed networks of activity is undertaken through the collection, analysis and deployment of various sorts of numerical and statistical information. This insight has been creatively applied in studies of the emergence of ethical trading initiatives and ethical audit (e.g., Hughes 2001). However, we look here at a different usage of calculative technologies by campaigning organizations. Freidberg (2004) notes that the development of ethical audit-ing initiatives, aimed at reconfiguring global supply chains, involves social movement organizations in efforts to generate and maintain media attention for their issues and causes. The production and dissemination of numerical information in the form of surveys, polls, charts and tables has become an important dimension of activist strategies (e.g., Grundy and Smith 2007; Hamilton 2009). It is this aspect of calculative technologies – as a campaign-ing device used to make claims and exercise influence in media publics and policy domains – that we focus upon. In this section, we look at two aspects of this process in ethical consumption campaigning: first the dissemination of alternative models of the consumer; and, second, the representation of this alternative consumer in various media publics.

The first aspect of representing the ethical consumer involves challenging the narrowly economistic model of the consumer as a self-interested egoist. From the perspective of purist economic liberalism, each person is seen as a sovereign actor determining their own conception of the good, and pursu-ing these by means of simple means–end rationality in the marketplace. In the UK context, the primary alternative to this 'neoliberal' discourse is associated with various 'Third-Way' understandings circulating around the New Labour governments of Tony Blair and Gordon Brown. It should be said that 'Third Way' understandings of the market and consumer choice differ significantly from the purist 'neoliberal' position (see Giddens 1998). They tend to circulate around the problem of pluralism: from this perspective there is no homogeneous sense of the social good or the public interest. It is this focus that leads to claims that 'the catch-all term "citizen" is unhelpful when it assumes there is a homogeneous "citizen interest"'.[20] These sorts of arguments are not invoked to support an unfettered individualism. Quite the contrary, what has come to be called the 'personalization' agenda is premised on the assumption that extending choice is the primary mecha-nism for ensuring that service providers will be responsive to the diverse needs of individuals and groups (see Pykett 2009). This perspective also entrains a particular understanding of 'democracy', one which privileges respecting people's preferences if these are properly informed choices, and assumes in turn that preferences are effectively expressed in the choices made in markets or surrogate markets. Consumer choice, in this 'market populist' paradigm, is a mechanism for reconciling the equally compelling concerns of individual 'aspiration' with pluralistic conceptions of the public good. In this paradigm then, people are understood less as 'citizens' responsible for the public interest, and rather as 'consumers, stakeholders or individuals concerned with the wider public interest'.[21]

Choice has, then, become a standard term in public debate in the UK, open to various interpretations. The limitations of the prevalent conceptualization of choice in public policy have become a focus of attention in a range of recent interventions by think-tanks and NGOs engaged in debates about these issues. While this problematization of consumer choice accepts certain precepts of the prevalent paradigm, it reinterprets them in ways that amount to a more thoroughgoing 'collectivization' of practices of consumer choice. We would locate the discourses of the 'ethical consumer' in the context of this broader problematization of choice. Here we see 'choice' being reconfigured as a dimension of civic engagement. In the process, the multiple motivations that are collected under the umbrella of 'choice' are unpacked:

> most people would support people's right to choose – if not on health principles, then on moral or efficiency ones.[22]

In practice, choice might be exercised on all three of these grounds – health, morality or efficiency – in the course of any simple activity, like the daily shop.[23]

What emerges from this field of alternative discourse is a figure of the 'citizenly consumer', actively choosing, indeed choosy, in the marketplace, but not necessarily on narrowly self-interested grounds at all. In this field, consumers are described with attributes usually associated with citizens. By way of example, let us return again to the Ethical Purchasing Index (EPI), which we discussed briefly in Chapter 2. The EPI is used to establish the size of the market in ethical goods and services, market share of ethical purchasing, and levels of growth in this sector. The key sectors measured by the EPI include fair trade, vegetarianism, organic foods, green household goods and responsible tourism. 'Ethical' is defined for the purposes of the EPI 'as personal consumption where a choice of product or service exists which supports a particular ethical issue – be it human rights, the environment or animal welfare'.[24] The EPI presents consumers as 'influential, proactive and engaged',[25] as supporting their communities by shopping locally, and as acting as citizens by rewarding companies with records of good practice.[26] The EPI is used to engage with a range of audiences: the general public, key retail stakeholders, policy makers and government departments. The EPI is both a 'catalogue' that measures ethical consumerism in order to lobby these actors, and thereby also a 'catalyst to its growth'.[27] The EPI is an example of an initiative that combines an emphasis on consumer choice with an argument for new forms of government regulation and for more proactive forms of corporate social responsibility. Consumer choice in a range of 'ethical' product markets is reinterpreted as an expression of broad public feeling in favour of certain sorts of collective goals that, on its own, consumer choice in the market cannot secure. Consumer choices therefore need to be empowered not only with 'information', but also by explicit intervention and endorsement by government in the form of regulatory interventions: consumers pull, producers push and governments endorse.[28] 'Choice' in the EPI is more than simply an aggregated market signal. It is discursively reframed as bearing other, more overtly political preferences.

Campaign organizations and think-tanks produce a variety of these typologies of the 'consumer' that when taken together are indicative of a broadly shared concern to better understand the diverse motivations that lay behind 'consumer' choice. In particular, there is an increasing concern to differentiate the 'ethical' motivations that shape consumer choice. For the Fairtrade Foundation, ethical consumers might be 'activists' (persuaders and supporters), or 'regular' ethical purchasers, or 'infrequent' ethical purchasers. For the Co-op, consumers might concentrate on 'looking after their own' or 'doing what I can'; they might be members of the 'brand generation', 'conscientious consumers', or 'global watchdogs'.[29] Business studies researchers are more blunt: ethical consumerism is divided between the 'die-hards' and the 'don't cares'.[30] And these exercises in categorization are not purely academic. They are put to work in the public realm to make visible the motivations that are hidden by thinking of consumer choice simply in terms of market signalling. They serve as the basis for various storylines through which 'ethical' dimensions of consumption practice – whether buying fair trade products, going 'organic', being more 'sustainable' – are publicly articulated in mediums from local newspapers to national television documentaries.

We have argued so far that 'choice' circulates in the public realm by being problematized, and that it is increasingly problematized in a register of *responsibility*. What we are suggesting in this section is that the discourse of consumer choice is open to re-inscription in terms which re-legitimize collective intervention in markets. We have already seen one version of this re-inscription – the 'New Labour' or 'Third Way' version in which choice is understood as a mechanism for ensuring more responsive modes of public service provision, conceptualized primarily in terms of principal–agent relations. In this discourse, the burden of ensuring that individual and collective outcomes are achieved is, indeed, thrown squarely on the consumer:

> If greater choice and control is extended to consumers, individuals must be prepared to take on more responsibility for the consequences of those choices.[31]

> [T]he public will be increasingly required to take responsibility for ensuring the public interest is balanced against individual needs.[32]

Another version of the re-inscription of 'consumer choice' is evident in the problematization of individual choice as bearing within it all kinds of 'risks', whereby rolling out mechanisms of choice to ensure more efficient service provision carries with it the likelihood that people will be allowed too much freedom to make bad choices. It is this concern that is evident in some of the interventions surrounding diet, obesity and smoking:

> our 'freedom' of choice is conditioned in newly unhelpful ways which misdirect our energies, and, as a result, individuals who make self-maximising choices often end up inadvertently minimising themselves instead. [. . .] The significance of prevailing value frameworks is heightened today by the fact that we are now being drawn to make choices that may not obviously impact on the freedoms of others or clearly injure the common good [. . .] but which are bad for us as individuals.[33]

Choice is reframed here as an inherently uncertain mechanism, just as likely to rebound on the individual as it is to undermine wider collective goals.[34]

It is on these grounds that a renewed justification of regulatory intervention to enable and enhance 'genuine' choice is developed. For example, the Fabian Society argues that there are numerous ways in which the same needs or wants can be met, depending on different 'choice sets'. Drawing on the ideas of Amartya Sen, a choice set is conceptualized as a collection of interconnected acts of consumption, the behaviour that comes with them and the production and infrastructure that supports them. Each choice set excludes or precludes other choices and options, so that 'there is no such thing as a purely "individual" act of choice: we always choose within a choice set'.[35] The argument is that individual rational choices do not necessarily lead to 'collective goods', as individual choices may circumvent or alter choices available to others. Here, then, we see a more explicit combination of discourses of individual responsibility with proactive arguments in favour of state and non-state intervention in the regulation and configuration of systems of provision.

This assertively 'citizenly' model of consumer choice forms part of a repertoire of narrative storylines mobilized by a range of organizations, including think-tanks such as the New Economics Foundation, the Fabian Society, the Food Ethics Council, Demos, Green Alliance, the Future Foundation; consumer groups such as the National Consumer Council; campaign groups such as the Ethical Consumer Research Association and the Fairtrade Foundation; and development charities such as Christian Aid and Oxfam. These organizations do not form a coherent movement. They have different organizational forms and membership bases, and they campaign on different issues, from public services to sustainability to global trade justice. Nonetheless, we can discern a family of related concerns around consumer choice and markets among this range of organizations. The idea that information is all that is required to ensure effective market supply in response to consumer demand for cleaner, fairer, greener products is increasingly being called into question across this discursive field. In debates around sustainable consumption, for example, choice is reconfigured in relation to 'institutional contexts'[36] and 'social scaffolding'[37]. Rather than either celebrating or resignedly accepting the triumph of consumer sovereignty, one finds various arguments that the key to effective change lies in providing infrastructures that support sustainable practices, combined with a degree of 'self-binding' constraint arrived at through regulating 'choice sets'. In short, the consumer-citizen is seen as a rational agent mobilized by information and educational devices *only* if these are accompanied by changes in the institutional settings and infrastructures of consumption. This reframing of choice and responsibility is typified by the 2006 report of the Sustainable Consumption Roundtable, an initiative of the National Consumer Council and the Sustainable Development Commission. Entitled *I will if you will*, the report argues that a 'critical mass' of citizens and businesses is waiting to act on the challenge of sustainability, but that it is constrained from doing so through lack of effective government support and direction.[38] The report is underwritten by the claim that expecting individuals or businesses to act 'sustainably' on the basis of isolated decisions is ineffective because neither set of actors has any sense of contributing to collective change. This example is indicative of a marked shift in thinking on sustainable consumption away from a focus only on the responsibilities of consumers. It emphasizes instead the proactive role of government in providing leadership and creating 'a supportive framework rather than exhorting individuals to go against the grain'.[39]

The emergent rationality of interventions which seek to problematize consumption is notable because it challenges the assumption that consumer choices in markets are equivalent to democratically expressed preferences that need necessarily to be respected. While some of the arguments made for state regulation are made on non-paternalistic grounds (i.e., in the name of the harms that certain patterns of individual choice bring about on other actors), what lies behind the discussions of *institutional contexts, choice sets* and *social scaffolding* is the claim that market choices are not necessarily a means of expressing preferences that deserve democratic respect at all. The rationality of calculative campaign devices such as the EPI supposes that consumer choices in the marketplace require supplementation, since they express deeper preferences that are only made visible through acts of additional interpretation. In this field of public debate, the meaning and significance of 'choice' is contested around an axis that holds that democratic governance should respond to and respect people's preferences. What is in question is just how to respond to what these preferences are; which preferences should be respected and which ones can be paternalistically substituted; and which actors are legitimately delegated to substitute their judgements on these matters for those of ordinary people (e.g., Thaler and Sunstein 2007). Arguments address the degree to which people have the 'volitional' will to make the choices that they ideally should prefer to make. The exemplary case of this type of justification for paternalistically preferring some form of substituted judgement for the expressed preferences of ordinary people is that of addiction (see Goodin 2002). And it is noteworthy in this respect just how much of the debate about responsible, sustainable and ethical consumption invokes a rhetoric of being 'locked-in' and 'addicted' to challenge narrow concepts of choice, information and preferences.

If the first aspect of representing the ethical consumer which we have discussed in this section involves reframing the economistic, egoistical models of the consumer, then the second aspect of representing the ethical consumer involves efforts to generate media attention for the causes for which this figure is a crucial support. Once again, the EPI provides our example of this process. In 2003, the EPI introduced new measures aimed at capturing the value of consumer behaviours that are not strictly based on the purchasing of specific items, such as spend on public transport, buying for re-use, local shopping, and avoidance or boycotting of 'unethical' brands. This reflexive adjustment indicates a concern that the EPI's measurable definition of 'ethical' goods and services might under-count the economic value of ethical consumer behaviour. And in the case of the criteria of local shopping and public transport at least, it also indicates the degree to which the exercise of 'choice' is shaped by systems of collective provisioning over which consumers have little direct influence. Survey data is often used to argue that, despite the existing size of these ethical markets, there remain various obstacles and blockages in the way of consumers translating their concerns into effective demand in the marketplace. The 2003 EPI Report acknowledged that, despite impressive signs of growth, the ethical market sector accounted for only 2% of total market share in the UK:

> Whilst ethical consumers can act as innovators in getting new products to the market, for real progress to be made supply side influences or government inter-vention may be required for some products to achieve mass market adoption.[40]

Here, the ethical consumer is invoked as an eager but frustrated subject, a potential but untapped market for retailers, and a potential partner in shifting market demand for regulators and policy makers.

The primary goal of the production of survey data like the EPI is the attraction of regular media attention. Media attention is a relatively low-cost resource for the sorts of campaign organizations involved in ethical consumption (see Gamson and Wolfsfeld 1993). One of the key objectives of ECRA, for example, is not simply to change consumer behaviour, but to raise awareness about a broad range of political issues through the medium of consumer policy and consumerism. This involves not only the publica-tion of specialist publications for supporters, which we have already dis-cussed, but also the garnering of regular news coverage in newspapers, radio and television. The annual publication of the EPI now gains regular news coverage, as do other similar survey-based research reports. For example, the Co-op's review of its retailing brand, *Shopping with Attitude*, received

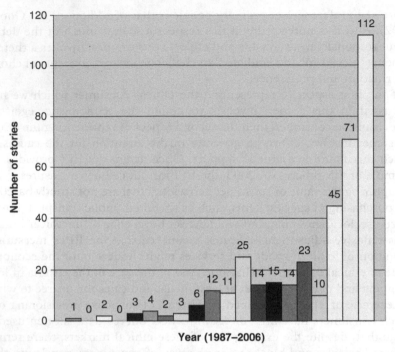

Figure 4.1 Increasing news coverage of ethical consumption between 1987 and 2006, as shown by related stories in UK national newspapers (*Source*: Nexis UK)

extensive news coverage in 2004, framed by headlines that identified a trend from 'Essex Man to Ethics Man' in the results of its survey of 30,000 people's concerns over safety, propensity to boycott goods and varying degrees of willingness to pay more for ethical products. Furthermore, ethical consumption issues have also breached that bastion of utilitarian consumer reflexivity, the Consumer Association's *Which?* magazine. Since 2003, *Which?* has carried regular items on aspects of ethical consumption – for example, on waste disposal and recycling, and on sustainable consumption. It has also begun to include 'ethical' inserts into its broader reviews of select products – for example, 'The ethics of the shoemakers' in its June 2003 report on running shoes and trainers, and the 'The ethics of making mobile phones' in its December 2003 review of mobile phones.

There has been a significant increase in the amount of regular news coverage of ethical consumption in the British news media over the past two decades. The first stories on this topic started to appear in the late 1980s, but the real increase has taken place since the mid-1990s (see Figure 4.1).

Two things stand out about this coverage. First, many of these stories depend on the types of information generated by surveys and opinion polls. Second, this increase in coverage is clearly related to the emergence of a select number of organizations as important and credible sources of news. ECRA is one of these organizations, so too is the Fairtrade Foundation and the Soil Association, whose annual Fairtrade Fortnight and Organic Food Week, respectively, also attract regular media coverage. These organizations are now 'certified' sources for stories on recycling, energy futures, sustainability, global labour rights and related topics of 'ethical' consumption.

For example, in 2004, the leading UK liberal daily paper, *The Guardian*, ran a year-long series in which one of its journalists, Leo Hickman, set out to apply 'ethical' principles to all aspects of his family's household consumption.[41] Two things are notable about this series, the most sustained mainstream media event around ethical consumption in the UK to date. First, the 'experiment' was supported by a set of auditors, from ECRA, The Soil Association and Friends of the Earth, who regularly advised on the specific issues at stake when it came to, for example, supermarket shopping or loft insulation. Second, the series explicitly focused on the difficulties involved in balancing competing demands to 'do the right thing' with the practicalities of everyday life, seeking to avoid a 'moralizing' tone in favour of raising dilemmas. This narrative focus on the dilemmas, difficulties and ironies of being 'ethical' is indicative of a broader, emergent rationality within ethical consumption campaigning that aims to engage with the range of people's existing dispositions, rather than one that simply preaches to abstracted individuals exercising consumer choice from the moral high ground.

Survey data and consumer information are an important aspect of the emergence of ethical consumption organizations as actors in the public realm, as well as of their more direct engagement with consumers and supporters. The growth of news coverage of ethical consumption issues is indicative of a successful alignment of the activities of campaigning organizations with the conventions and imperatives of professional news production, so that by producing the type of information resource that news organizations *need* – survey data and opinion polls on consumer preferences – these organizations can establish their own *value* as credible sources (Gamson and Wolfsfeld 1993). The steady growth of news coverage about ethical consumption indicates two features of the organized mobilization and articulation of 'the ethical consumer' in the public realm in the UK. First, it reflects a successful strategy by organizations active in this area in amplifying their cause through media-led repertoires (Freidberg 2004). And, second, it underscores the extent to which these organizations are primarily involved in *brokering* various sorts of information and expertise amongst different actors, whether these are news organizations, retailers and suppliers, or ordinary consumers.

The two aspects of the politics of information around ethical consumption discussed in this section – making visible and speaking for 'consumers' in the public realm – throws into new perspective the strategies discussed in section 4.3. If one looks only at the provision of information to consumers discussed in section 4.3, and if one ignores the self-selecting quality of the audience for this type of information, it is easy to conclude that this is complicit with a broader privatization of responsibility, now articulated through the aggregated preferences of sovereign consumers. And, of course, understood in these terms, a straightforward critique of ethical consumption suggests itself. Looked at in purely economic terms, the impact of ethical consumption remains only a pinprick on unequal patterns of world trade or the corporate domination of domestic retailing.

The analysis in this section of the two-fold articulation of the 'ethical consumer' suggests that this type of critique only sees half the story. By factoring in the other aspect of the deployment of information – the ways in which information about consumers enables organizations to speak for the ethical consumer as a concerned citizen of the world – a more complex articulation of individual action and collective organization emerges. For the organizations campaigning around the growth of ethical consumption in the UK, consumer-based activism is an important way of raising awareness about issues and establishing the legitimacy of their own claims and the validity of their own arguments. In the UK, organizations such as Traidcraft, the Fairtrade Foundation, Oxfam, Christian Aid or the Co-operative Group are all active in trying to exert influence over governments and corporations regarding issues of Third World debt, trade justice, corporate social responsibility and international human rights. Their capacity to act in this way in networks of transnational political advocacy depends on being able to show that they have broad-based popular support for the sorts of changes that they are promoting. A basic requirement for any organization involved in this sort of activism and advocacy is to sustain a constant public presence by demonstrating the *number* of its supporters and the *intensity* of their commitment (Tilly 1994). In this light, and given the notorious difficulty of mobilizing consumers as political subjects even around 'consumer' issues, using surveys and polls to demonstrate a growth in sales of fairly traded products, organic food or ethical investment is a relatively low-cost strategy available to organizations for performing their legitimacy in the wider public realm, as well as validating themselves to members and supporters.

In this chapter, we have examined the process by which ethical consumption campaigning uses various discursive devices – information-rich web sites, magazines, surveys and polls – to assemble the disparate practices of anonymous consumers into coherent indices of 'ethical' preferences in the effort to exert normative force over public and private actors. We have questioned whether the primary objective of these campaigns is to create new 'subjects' from scratch. We have argued instead that these campaigns seek to channel existing but disparate dispositions into focused engagements with state agencies, corporations, or regulators around specific issues. Ethical consumption campaigning is characterized by specific rationalities and strategies through which various organizations from outside the realm of formal politics seek to 'act upon the actions' of ordinary people at the same time as they seek to articulate these actions into networks of affiliation, lobbying and mobilization that address powerful and often global actors. By examining the means through which organizations both speak of and speak for 'ethical consumers' in the public realm, this chapter has demonstrated that it is acts, not identities or subjectivities, which matter in mobilizing the presence of 'ethical consumers' in the public realm – acts which can be measured, reported, calculated and represented in the public realm.

4.5 Conclusion

The contemporary politicization of consumption has a double aspect: consumption is constructed as an object of concerted political action at the same time as consumption is increasingly framed as a medium for the mobilization of support and activism by campaign organizations. In this chapter, we have suggested that ethical consumption campaigning redefines everyday consumption as a realm through which consumers can express a wide range of concerns and engage in a broad set of projects, including social justice, human rights, development or environmental sustainability. It does so by circulating narratives in media publics, large and small, in which everyday life is reframed around a series of dilemmas concerning what people ought to do. These dilemmas revolve around *actions* – what to buy, where to invest, whether to drive or walk, what not to buy, whether to help – and *issues* – helping the global poor, reducing carbon footprints, supporting trade justice.

We have suggested in this chapter that ethical consumption campaigning is actively involved in 'globalizing the consumer'. Ethical consumption is a movement distinguished by advocacy-type organizations that specialize in the production and dissemination of information, knowledge and narrative storylines, and which are embedded in transnational networks of labour solidarity, environmental advocacy, trade justice and related issues. Deploying these resources, these organizations endeavour to articulate *consumption* and *the consumer* through a register of 'ethics' and 'responsibility' that seeks to configure people as political actors embedded in networks of global action. But it is important to note that the political rationality of ethical consumption campaigning does not either aim for or require the complete overhaul of people's identities as 'ethical consumers'; it aims to be responsive to emergent dispositions and structures of feeling, translating these into forms of collective, concerted action (Soper 2004). If ethical consumption campaigning has any effect in producing new actors in the public realm, this need not take the form of fully formed, embodied elaborations of the self at all. It is more properly thought of in terms of the production of various *singularities* – a purchase, an investment, a donation – that can be registered, recorded and reiterated through other circuits of communication.

Understood as a broad-based movement of organizations and supporters, ethical consumption campaigning can be seen as contributing to the reconstruction of political responsibility in an unequal world discussed in Chapter 1. Ethical consumption can be seen as an example of a style of political practice in which various citizenly acts are undertaken through the medium of everyday practices such as shopping, travelling to and from work, or disposing of household waste. The growth of ethical consumption involves the production and dissemination by various agencies of a set of 'moral risks' that people are told they face as consumers – the risk of being implicated in some way in the reproduction of harm to other people, or to the environment, or to future generations. The actors involved in campaigning around ethical consumption are therefore certainly engaged in the moralization of consumption (Miller 2001c; Hilton 2004). But this moralization does not simply dismiss consumption as individualistic, acquisitive and self-interested. Rather, it reframes consumption in terms of the collective responsibilities that people are implicated in by virtue of their status as consumers. 'Being ethical' is understood in particular ways in this set of practices – in terms of avoiding or diminishing one's implication in the reproduction of harms, for example, and along broadly consequentalist lines that anchor 'responsibility' firmly around an analysis of the intended and unintended consequences of one's own actions.

We saw in Chapter 3 that it is common to argue that power in the contemporary world is now diffused among consumers (e.g., Miller 1995; Amin and Thrift 2005). Our argument here is that the exercise of this potential in the public realm depends on the purposeful organization and articulation undertaken by social movement organizations, non-governmental actors and activist networks. The growth of ethical consumption campaigning represents a development within the repertoires and strategies of social movements and NGOs more generally. This reflects the strategic choices made by organizations and activist groups to mobilize 'the consumer' in particular ways, faced with various opportunity structures and the availability of different bundles of resources. In Chapter 3, we argued that the problematization of expanded commodity consumption and the explicit mobilization of 'consumer' identities are only contingently related. In this chapter, we have argued that the mobilization of the consumer as an 'ethical' subject, enrolled into various collective projects of solidarity, is best explained with reference to the strategic repertoires deployed in contemporary activist and advocacy politics. Effective activist communication has increasingly adopted a lifestyle vocabulary, anchored in consumer choice, self-image and personal displays of social responsibility (Bennett 2004). Consumer-oriented forms of activism have, then, become modular across different issues and movements. This internal shift in activist repertoires is in turn connected to the emergence of an external political environment in which the rhetoric of globalization, free markets and consumer choice can be critically re-inscribed to provide new storylines to potential supporters, at a time when forms of political contention are increasingly articulated across national boundaries through various networked spaces.

Our aim in this chapter has been to establish that what in the UK, at least, is routinely referred to as ethical consumption is a *political* phenomenon. It is a phenomenon in which the registers of 'ethics' and 'responsibility' are deployed in pursuit of some classically political objectives: collective mobilization, lobbying, and claims-making around issues of broad public concern, in pursuit of equality, fairness and justice. There is no reason to suppose that there is a zero-sum relationship between deploying the narratives and devices of consumerism as a surface of mobilization and other, more conventionally political, modes of action. Quite the contrary: the repertoires of consumerism are a means of extending existing dispositions into new areas of practice, and are related to new forms of public action by organizations concerned with a range of contentious issues. In order to better understand ethical consumption campaigning as political action, we need to break with the assumption that these activities aim primarily at addressing people as rational economic actors through the medium of information. In the next chapter, we develop further the understanding developed in this chapter of the ways in which campaigns provide storylines to argumentative subjects faced with an ongoing set of everyday dilemmas around 'doing the right thing'. We now turn to an examination of the forms of routine reasoning that people who find themselves positioned as 'consumers' engage in when confronted with a proliferating range of potential acts of *responsible choice*.

Part Two

Doing Consumption Differently

Chapter Five

Grammars of Responsibility

JOHN: *Everyone says they were happy 20 years ago.*
KAREN: *I think in some ways though, life was just simpler. I think all this choice and stuff . . .*
ARUN: *. . . has just complicated things.*[1]

5.1   Justifying Practices

We have argued in previous chapters that a defining feature of the contemporary politicization of consumption is the deployment of a vocabulary of responsibility to problematize various individual and social practices, from driving to work, going to the doctor, investing for the future, going on holiday, feeding the kids, or just eating too much chocolate. The prevalence of this vocabulary of responsibility suggests that everyday consumption practices are being publicly redefined as ethical practices, in the sense that injunctions about what *one ought to do* are combined with strong appeals to people's sense of personal integrity and sense of self. In short, as we argued in Chapter 2, consumption has become a surface for the 'ethical problematization' of the self. By ethical problematization, we refer to the myriad practices which construct 'those things that our moralities might engage with' (Hodges 2002: 457). In this chapter, we want to further pursue this understanding in order to develop an alternative to the three prevalent approaches to understanding the ways in which ordinary people are implicated in the politics of consumption: first, as creative appropriators of commodities involved in identity-formation; second, as interpellated subjects of neoliberalism and/or strategies of de-fetishization; or, third, as attached to commodities in deeply ingrained, affectively charged ways. The concept of

Globalizing Responsibility: The Political Rationalities of Ethical Consumption by Clive Barnett, Paul Cloke, Nick Clarke and Alice Malpass © 2011 Clive Barnett, Paul Cloke, Nick Clarke and Alice Malpass

ethical problematization leads us to focus instead on the narratives through which selves are constituted and maintained through time.

Approaching the emergence of initiatives around ethical consumption in terms of ethical problematization suggests a methodological entry point that focuses on processes of discursive positioning and rhetorical, argumentative practices. We have argued that the political rationalities of ethical consumption campaigning seek to incite individuals to reflect and deliberate about everyday routines. Understanding how these incitements are negotiated requires us to reconstruct the interactive contexts in which individuals argue, justify, and give accounts of themselves. It is these interactive contexts which standard post-structuralist views of subject-formation have difficulty accounting for, either theoretically or empirically. In this chapter, we explore the theoretical and empirical issues raised by conceptualizing ethical consumption in ways which do justice to people's own competencies as persons and not just 'subjects'.

This chapter develops the concept of ethical problematization in order to bypass the restriction of our understanding of subjectivity to two mutually supportive alternative visions: an emphasis on 'ideological interpellation' and 'subject disciplining' on the one hand, and a more recent emphasis on the 'sociality of affective backgrounds' (e.g., Clough 2009). Between them, these two approaches – one that sees subjectivity as an effect of strategies of power working through subjection, and one that sees subjectivity as an effect of iterative, habitual affective routines – elide what is most distinctive about thinking seriously about discourse. While 'non-representational' approaches dismiss discourse as irredeemably 'representational', the interpellative imagination constructs discourse as a medium of top-down subjection of individuals. Both approaches share a continuing trace of mechanistic conceptions of subjectivity, either as an effect of interpellative address and recognition, or as an effect of strategic interventions which deploy affective means of manipulation. And both approaches erase from view the distinctive pragmatic lineage of discourse as *a concept of action* (Potter *et al.* 2002).

Our argument in the preceding three chapters suggests that attending to the rationalities of ethical consumption campaigning helps us to see a way beyond these conceptual problems. In contrast to the view that hegemonic neoliberalism seeks to roll out coherent subject-effects, we have suggested that far from being straightforwardly championed and promoted, 'consumer choice' is circulated as a term in policy discourse and public debate by being problematized. This means that far from inviting individuals to recognize themselves as consumers in the circular scene of interpellative address, contemporary discourses of consumerism generate dilemmas and incite people to reflect on their habits and routines. In short, they invite them to engage in the reflexive monitoring of their conduct (Giddens 1984). And in contrast to the 'non-representational' elision of cognition into so many 'more-than-rational' affective triggers, these sorts of strategic interventions in the field of consumer agency also presuppose an ordinary, embodied capacity for articulating background. In short, for both conceptual and empirical reasons, restricting ourselves to the choice between an 'interpellative' imagination of subjectivity and a non-representational imagination of 'affect' leads to the same impasse. Both approaches close down analysis of the ordinary forms of normative reasoning woven into practices. Sayer captures what is most problematic about these sorts of approaches:

> Despite the centrality of what we care about to our lives, much of social science abstracts from this ordinary phenomenon, presenting us with bloodless figures who seemingly drift through life, behaving in ways which bear the marks of their social position and relations of wider discourses, disciplining themselves only because it is required of them, but as if nothing mattered to them. (Sayer 2005: 51).

Both strategic relations of power, the key to interpellative models of subject-formation, and habitual action, the key to non-representational approaches, are important dimensions of social relations. But neither separately nor combined can they be considered to exhaust what is involved in ongoing practices. Most importantly, neither gives adequate weight to what Sayer (2005: 6) calls 'lay normativity'. This concept restores to view the 'range of normative rationales, which matter greatly to actors, as they are implicated in their commitments, identities and ways of life. Those rationales concern what is of value, how to live, what is worth striving for and what is not'. It is precisely these lay normativities that, as we have argued already in previous chapters, are targeted by ethical consumption campaigning, and therefore it is incumbent on research on this phenomenon to develop theoretical and empirical strategies which factor in the dynamics of these lay normativities.

Taking lay normativity seriously is not, it should be said, merely an 'empirical' matter of looking at the 'reception' of top-down discourses by subjects. It points to a fundamental social-theoretical commitment to acknowledging the importance of practices of justification and accountability as constitutive of social coordination and human relations (e.g., Boltanski and Thévenot 2006; Flyvbjerg 2001; Tilly 2008). Taking justification and accountability seriously as a constitutive feature of social coordination requires us to rehabilitate the concept of discourse. In this chapter, we take discourse seriously as a methodological entry point and as a factor in social practices, and in so doing we revisit the concept of *positioning* that is central to discursive theories of self-formation.

Of course, in generic poststructuralism, the positioning of subjects in discourse is widely understood as the primary mechanism for the inscription of hegemonic identities into the habits, feelings, identifications and conduct of living individuals. As we argued in Chapter 3, restoring to view the role of interpretation – that is, of judgements and justifications – in practices suggests an alternative understanding of *positioning* in processes of subject-formation.

This understanding needs to acknowledge the interactive, inter-subjective, communicative dynamics through which norms are iteratively performed:

> the power of discourses could only be realized if individual human subjects were prepared to engage with them, positioning themselves with respect to the discourses, with consequences for their selves and identities. (Day Sclater 2005: 323).

Understanding the concepts of positioning, subjectivity and discourse with reference to practices of justification and accountability is quite consistent with a broad tradition of 'non-representational' thinking. It builds in particular on the work of John Shotter (1993a, 1993b), Rom Harré (1991, 2002, 2004), and other 'discursive psychologists' in adopting a rhetorical conception of the person, in which subjectivity is 'responsively relational' all the way down. This tradition restores discourse to its action-oriented sense, understood as a 'domain of accountability'. Restoring to view this sense of discourse as a practice will allow us to develop the thought that the human subject should be understood primarily as 'socially constructing' rather than 'socially constructed' (Thrift 1996: 129). And we emphasize that this is a recovery, not a new understanding. There is no need to think of discourse as practice *for the first time*, in the wake of recent non-representational and materialist turns. This understanding is already there in the genealogy of this concept. Rather than thinking of the self or subjectivity as being 'constructed' through vertical positionings, that is, as an effect, we might be better served to focus on self-making as a dynamic process of being made and making (Day Sclater 2005). Thinking of discourse as a concept oriented to action (Potter and Wetherell 1990; Potter *et al.* 2002) leads methodologically to a focus on talk-data not as a mirror on the mind, but as a resource for understanding the ways in which ordinary reflective, deliberative capacities of subjects-in-practice fold together habit and reflection.

In the rest of this chapter, we develop an analytical approach to discourse and talk-data that turns the tables on prevalent models of discourse analysis. Rather than tracing those moments when subjects are positioned by hegemonic discourses, we focus on *the ways in which people attribute meaning to the attempts to position them as subjects of various ethical responsibilities that can and should be discharged through the medium of everyday consumption.* In short, we aim to do justice to the *grammars of responsibility* through which people reflect on the cares, concerns and duties of everyday consumption practices. By 'grammars of responsibility', we mean to point to the need to discern just what actions are being performed when informants have recourse to 'responsibility-talk'. We use 'grammar' here in a Wittgensteinian sense, to refer to the idea that analysis can focus on the norms through which meaningful statements show up in the world as certain sorts of action. As Wittgenstein (1953: §372) put it, 'Grammar tells us what kind of object anything is'. Assertoric propositions are not always, if ever, straightforward reports of matters of fact, whether about the world or an interlocutors beliefs. They are expressions of normative attitudes to states of affairs and relationships in which subjects are located. In what follows, our argument is that faced with moral demands about what to do with respect to some aspect of their practices which involve consumption, people sometimes make *excuses* for not doing what is presented as the 'ethical' thing to do; but sometimes they make pertinent sounding *justifications* for not considering these demands as relevant or binding on them at all; and sometimes they articulate justifiable scepticism towards the whole frame of 'responsibility' that is being addressed to them as individuals (see Austin (1961) on the distinction between excuses and justifications). Remaining alert to the 'grammar' of responsibility-talk is to be concerned with spotting the difference between these actions when they are performed. It is, in short, a matter of trying to focus in on 'what happens when people in their own worlds start giving, receiving, and negotiating reasons' (Tilly 2006: 31).

Section 5.2 of this chapter reviews the prevalent ways in which ordinary talk about consumption is used methodologically in studies of ethical consumption, before developing a theoretical argument for why discursive theories of positioning provide an alternative entry point for understanding the dynamics of ethical consumption campaigning and everyday practices. Sections 5.3 and 5.4 analyse talk-data from our focus-group research undertaken in and around Bristol during 2003 and 2004, in order to develop an understanding of how ordinary people engage with the moral demands circulated by ethical consumption campaigns. Section 5.3 investigates the *versions* of reality through which people ascribe meaning to these demands. Section 5.4 investigates the ways in which they engage in practical reasoning which transforms these demands into *dilemmas*.

5.2   Researching the (Ir)responsible Consumer

Research on how individuals relate to ethical consumption policies and campaigns falls into three broad strands. (See Crane (1999) and Newholm and Shaw (2007) for critical reviews.) The first approach, one prevalent in business studies, management studies and economics as well as strands of sociological research, involves survey-based research on individuals' *attitudes* to various consumption-based issues (e.g., Shaw and Newholm 2002; Shaw and Shui 2003). These research data are routinely framed in terms of the question of whether consumers' buying behaviour is consistent with the expressed attitudes to ethical products (e.g., De Pelsmacker *et al.* 2005; Clavin and Lewis 2005). A common refrain amongst academics, policy makers and campaigners is that this type of research shows that there is a gap between attitudes and behaviour – that avowed support for various 'ethical' causes is higher than practical action in support of these causes. Discovering attitude/behaviour 'gaps' depends on the reification of atti-tudes and opinions as pre-existing attributes of individuals, assuming that they can be uncovered by social science survey methods (Myers 1998; Wiggins and Potter 2003). Rather than assuming that individuals have 'attitudes', internally located in their minds, which may or may not be expressed externally in 'behaviour', it is better to think of attitudes rhetori-cally, as expressions made in contexts where argument and evaluation is likely: 'attitudes are stances on matters of controversy' (Billig 1991: 143). Expressing attitudes, in short, is a form of behaviour, and attitudes should be approached analytically as forms of evaluative discourse, not as inde-pendent states of mind but as positions taken up in relation to dilemmas: 'Attitudes, far from being mysterious inner events, are constituted within the business of justification and criticism' (ibid.).

A second strand of social science research on ethical consumption, and in particular on sustainable consumption initiatives, adopts a more qualitative perspective that allows that descriptive versions of peoples' attitudes and opinions are evaluative. Such work acknowledges that individuals deploy various discourses of responsibility in relation to policy initiatives aimed at provoking changes in everyday consumption practices (e.g., Hinchliffe 1996, 1997). There is an assumption in much of this research that the 'vocabularies of motive' deployed in research situations where informants discuss issues of environmental risk and sustainability are primarily dis-courses of *blame*, used to displace responsibility from the individual to governments, private businesses or 'society' (e.g., Bickerstaff and Walker 2002; cf. Jackson *et al.* 2009).

Research in this second strand has developed into a third strand of research, again most developed in debates about sustainable consumption. This research often reiterates the argument that public attitudes on envi-ronmental issues tend to run ahead of behaviour, but reframe this 'gap' by acknowledging that people are often 'locked in' to particular patterns of consumption. Tim Jackson (2003, 2004), for example, argues that research on sustainable consumption initiatives needs to acknowledge the impor-tant symbolic role that consumer goods play in social relations (cf. Belk 1988), and that this makes behaviour change much more complex than is often assumed by policy paradigms that focus on the provision of information about environmental costs and risks. Without acknowledging the social function of consumption, simple appeals that people should reduce or change their consumption activity 'is tantamount to demanding that we give up certain key capabilities and freedoms as social beings' (Jackson 2003: 9). Jackson's reconceptualization of the social psychologies of sus-tainable consumption practices is part of a broader critical reassessment of established paradigms in this field, one focused on developing more sophis-ticated understandings of the psychological investments that surround everyday consumption activity (e.g., Barr 2003; Barr and Gilg 2006; Hobson 2006b). Qualitative findings which show that individuals respond sceptically or negatively to policy initiatives aimed at getting them to change their consumption activities are interpreted as providing important lessons about the limits of those policy strategies (e.g., Hobson 2002; Macnaghten and Jacobs 1997; Myers and Macnaghten 1998). This research has in turn helped to generate an appreciation of the extent to which changing consumption activities requires a focus on finding ways of integrating new consumption habits into the routines of everyday, domestic, often embodied routines (Marres 2009; Macnaghten 2003; Slocum 2004).

The third strand of research overlaps with the practice-based conceptualization discussed in Chapter 3. However, there remains a strong element of prescription in this type of social science, due no doubt to the extent to which this research is closely tied to specific fields of environmental and sustainability policy. Even as understandings of the social dynamics of consumption become more sophisticated, there remains a strong presumption that the attachments and investments that people have with patterns of consumption, as well as their doubts and scepticism, are *obstacles* or *barriers* to be overcome. This strand of research remains caught between the idea that providing information to individual consumers is a way of enabling them to act on their own preferences for more responsible futures, and the idea that changing consumer behaviour might require more than just providing lots of information. This impasse reflects the deep assumptions of a research field that continues to conceptualize 'the political' primarily as a realm of policy, regulation and legitimation, rather than one of mobilization, participation and contestation. A great deal of research on contemporary consumption focuses on questions of *whose* responsibility it should be to act to reduce harmful patterns of behaviour. Are the key agents of change consumers, or governments, or business, or the media, or NGOs, or professional bodies or religious bodies?[2] What policy- and governance-oriented research seems unable to acknowledge – unable to hear – is the degree to which their research subjects are able to articulate sceptical questions about just *whose definition of responsibility* has come to dominate public discussion and insinuate itself into their own practices through diverse mediums of the ethical problematization of everyday consumption. One aim of this chapter is to listen more carefully to this sceptical aspect of people's ordinary reasoning about the ethics of their own consumption activity.

In contrast to these three strands of research, in this chapter we seek to discern just what actions are being performed when informants provide evaluative accounts of various publicly circulated discourses of responsibility. We follow Gregson and Crewe (2003: 12) in recognizing that 'consumers actually know a very great deal about what they do by way of consumption, and that they can articulate this discursively'. We also assume that it is possible to take such discourse seriously as throwing light upon everyday practices without assuming that such talk is either a mirror of the internal mind or an accurate account of action (ibid.). As we argued in Chapter 3, talk and action need to be understood as reflexively related in practices, and this assumption directs our analytical strategy in this chapter. In order to further develop this perspective, we need to reconsider the action-oriented genealogy of the concept of discourse in social theory. In particular, we want to avoid the representational recuperation of the concept of discourse that is prevalent in generic poststructuralist theories of the disciplinary interpellation of subjects, a construal in which discourse is offset against 'materiality'. In this formulation, discourse is understood as the medium through which hegemonic programmes secure themselves at the level of subjectivity, by inscribing individuals into certain subject-positions. This is an understanding that underwrites poststructuralist theories of subjectivity across a broad strand of work, from post-Althusserian approaches to hegemony-as-discourse to 'Butlerian' accounts of performativity. This understanding depends upon elevating a semiotic understanding of discourse over a pragmatic understanding of discourse (cf. Fraser 1997: 151–170). It is formalized in 'critical discourse theory', which splices this understanding together with a model of discourse as a medium of 'power' derived from Foucault to develop a methodology for reading power-effects through various representational mediums (see Hammersley 1997).

We prefer to develop the line of thought that thinks of discourse as a concept related to practices of rhetoric. This perspective draws on a tradition of thought that includes J. L. Austin, Mikhail Bakhtin, Kenneth Burke, Rom Harré, Paul Ricoeur, Lev Vygotsky, Ludwig Wittgenstein and others (Wetherell *et al.* 2001b). It is an approach that redeems the concept of discourse as oriented to action, as language-in-use; in short, discourse is a *verb* (Potter and Wetherell 1990). This alternative understanding is most well developed in certain strands of environmental policy analysis (Hajer 1995: 42–72), and in discursive psychology (Potter 2003). It informs what Shotter (1993b) calls a responsively-relational understanding of subjectivity.

A key feature of this redemption of the concept of discourse is a reconfiguration of the idea of the 'positioning' of individuals as subjects in 'discourses'. The concept of position is central to the notion of the discursive construction of selves which understands talk as a form of action rather than an expression of roles, interests or attitudes (Harré and van Langenhove 1991: 393–394). From this theoretical perspective, discursive positioning is a form of joint action:

> the discursive process whereby selves are located in conversations as observably and subjectively coherent participants in jointly produced story lines. There can be *interactive positioning* in which what one person says positions another. And there can be *reflexive positioning* in which one positions oneself. However it would be a mistake to assume that, in either case, positioning is

necessarily intentional. One lives one's life in terms of its ongoingly produced self, whoever might be responsible for its production. (Davies and Harré 1990: 48, emphasis added)

On this understanding, positioning is a central organizing concept for analysing 'how it is that people do being a person' (ibid.). It implies a distinctive understanding of the relationship between people and their conversations:

> A subject position is a possibility in known forms of talk; position is what is created in and through talk as the speakers and hearers take themselves up as persons. (Ibid.: 62)

In positioning theory, agency emerges from the negotiation of constraint and choice in ongoing discursive activity. This understanding of the discursive construction of the self through positioning informs an avowedly 'non-representational' tradition of discursive psychology, one in which analysis of 'mind' is no longer framed through internal/external dualisms, and therefore one in which divisions between social/psychic, practice/discourse are reconfigured in terms of ongoing streams of action. Positioning is a concept that 'offers a dynamic, agentive model of identity construction where a person creates a possible identity for themselves in a particular context through their active positioning in relation to, or perhaps in opposition to, elements in their discursive cultural context' (Linehan and McCarthy 2001: 436).

One limitation of this approach is the sense that the self exists only in fleeting identifications with discursive positions. The problem of how to account for the persistence of the self over time leads advocates of this approach to embrace a narrative understanding of the self, in which identifications acquire meaning by being associated with biographical plots and storylines. From this perspective, identities are maintained by integrating events into coherent narratives; this narrative understanding provides an answer to the question of how and why individuals invest or dissent from certain subject positions (see S. Taylor 2005; Taylor and Littlejohn 2006).

The development of this tradition of discourse theory is shaped by a recurrent impasse, one that reflects different disciplinary traditions of social psychology and conversation analysis (Wetherell 1998; Hammersley 2003). Conversation analysis tends to favour a sense of positioning as a process of turn-taking between participants in ongoing talk – as a kind of horizontal argumentative give-and-take. Discursive psychologists and critical discourse theorists tend to understand positioning as a mechanism for the reproduction of hegemony – as a more determinant, vertical framing of possibilities which provides various 'interpretative repertoires' in which individuals struggle to position themselves. In order to negotiate this impasse empirically, our own research used focus groups to generate talk-data that elaborates two dimensions of positioning. First, talk-in-interaction involves 'horizontal positioning', in so far as focus groups are effective in provoking interactive conversations amongst participants; second, talk-in-interaction involves 'vertical positioning' in so far as the facilitators of the focus groups framed the conversations around various explicitly articulated questions and topics. In attending to both dimensions of positioning, we are seeking to make a virtue out of the observation that social science research itself involves positioning of subjects in discourse (see Harré and van Langenhove 1991: 404–406). Focus group data can be understood as a resource through which to understand the forms of relational reasoning that people are able to bring to bear on the various moral demands addressed to them by ethical consumption campaigning.

In presenting focus group research in this way, as evidence of how people negotiate horizontal and vertical positionings in discourse, we are following a strand of academic theory that understands focus groups as useful in generating evidence of interactional processes (see Cloke *et al.* 2004; Kitzinger 1994; Wilkinson 1998, 2006). In focus groups, the researcher invites participants to reflect on problems which might have personal relevance to them; and participants are free to respond with answers of their own, in interaction with other participants, and are free to bring up events or topics of their own (Hermans 2001). Focus groups generate two sorts of interaction: complementary interactions, in which people share common experience and agreed perspectives; and argumentative interactions, in which participants' question, challenge and disagree with one another (Kitzinger 1994: 107). A key point about this perspective is that the effect that the expression of opinions by one participant has on other participants is not considered a problem since focus groups are not primarily aimed at eliciting individual perspectives. In recommending this approach, we are concerned with generating evidence of how ordinary people engage with publicly circulated moral discourses of responsible consumption by engaging in inter-subjectively situated, evaluative interpretations of those discourses:

> Focus groups have been used as a way of finding opinions and underlying attitudes, but they can also lead us to reflect critically on what opinions are, and what people do with them. Focus groups can be seen as experiments in constituting a public domain, one that is new for many of the participants.
> (Myers 1998: 106)

Focus groups can, then, certainly be understood as 'a convenient way of obtaining a lot of immediately relevant on-topic talk' (Edwards and Stokoe 2004: 505). But they are particularly valuable in generating evidence of the co-construction of meaning (Wilkinson 1998). For this reason, there is no need to assume that this methodology can support inferences about attitudinal consensus in groups (cf. Burgess *et al.* 1988). Any consensus is best thought of not as reflecting pre-existing attitudes, but as emerging from the dynamics of interaction. Focus groups are well-suited to theoretical generalization, and it is this that attracts us to this methodology. Our argument here is focused on developing a theoretical understanding of the interpretative mediation of public discourses of ethical consumption into contexts of everyday life. The approach to talk-in-interaction developed in this chapter is, then, geared towards the broader aims of this book, according to which any understanding of how the exacting and contradictory moral imperatives of ethical consumption discourses are made sense of by ordinary people needs to take seriously 'what matters' to people.

One strand of focus group research which draws on theories of discourse privileges the concept of interpretative repertoire, holding that analysis should focus on identifying the publicly available discourses drawn on in interaction, and assessing the dispositions adopted by participants to these repertoires (e.g., Jackson *et al.* 1999). The concept of interpretative repertoire, referring to a coherent set of shared themes and tropes drawn upon in evaluative talk, is a means by which discursive methodologies figure questions of 'power' into their analyses. The concept of interpretative repertoire brings into view the impasse in how the 'social' qualities of discourse are understood in avowedly discursive theories of the self – an impasse between analysing talk as a form of joint action, or analysing talk as expressing publicly available, socially differentiated repertoires. Wetherell (1998) suggests a synthetic approach, which combines the emphasis on the active taking up of positions in interaction derived from conversational analysis with the stronger sense of the structuration of positioning found in poststructuralist approaches. As Wetherell (ibid.: 402) puts it, poststructuralist approaches rarely focus on the dynamics of interaction, while conversation analysts 'rarely raise their eyes from the next turn in the conversation'. In her synthesis, there are two levels of positioning going on in any episode of interactive talk, both of which depart from individualistic understandings of the self. First, there is positioning by previous turns in a conversation, implying a strongly collaborative account of the production of selves in interaction, involving both 'interactive positioning' by others and 'reflexive positioning' of oneself. It is this level that we have dubbed *horizontal positioning*. Second, there is positioning by hegemonic discourses, a more 'macro-level' and externalist perspective. This is what we refer to as *vertical positioning*.

In the next two sections of this chapter, we build on the theoretical and methodological perspective on discursive positioning developed in this section to analyse how people from a range of social backgrounds attribute meaning to publicly circulating discourses that attempt to position them as *consumers* with *responsibilities* to various global issues. Thinking of focus groups as occasions for investigating the dynamics of horizontal and vertical positioning, in section 5.3 we endeavour to clarify the *grammar* of responsibility-talk provoked by reflecting on the ethics of consumption – to clarify what actions people are performing when they deploy responsibility-talk of various sorts. Drawing on the concept of 'factual versions', in this section we seek to do justice to the ways in which our informants interpret 'position calls' by placing them under certain descriptions of reality. In section 5.4, we move on to consider how our informants critically engage these position calls through forms of reasoning which transform moral demands into practical dilemmas.

Taken together, the next two sections aim to elaborate on the routine forms of *practical reasoning* available to ordinary people when confronted with publicly circulated demands to consume 'ethically'. We take practical reasoning to refer to 'reasoning which we and others can use both in personal and public life not merely to judge and appraise what is going on, not merely to assess what has been done, but to guide activity. The activities to be guided range from institution building and reform to the daily acts and attitudes of personal life' (O'Neill 1996: 2). There are, then, two aspects of practical reasoning: an orientation to ensuring that ways of guiding action are *practical*, 'in that they can help agents with quite limited and determinate capacities to live their lives' (ibid.); and an orientation to ensuring that they are *reasoned*, in the sense that they refer to some sort of justifiable authority. We take practical reasoning to be an ordinary practice constitutive of ongoing social coordination. In section 5.3 we elaborate on the ways people reflect on the practical imperative; and in section 5.4, we elaborate on their accounts of the imperative for actions to be reasoned.

5.3   Versions of Responsibility

The empirical analysis presented here draws primarily on data generated in ten focus groups undertaken from the Autumn of 2003 to Spring 2004.[3] Focus groups were recruited in ten different wards of Bristol, in a range including the least to the most deprived wards in the city: Ashley, Bishopston, Easton, Hartcliffe, Henleaze, Knowle, Southmead, Southville, Stockwell and Windmill Hill (see Figure 5.1).

Participants in these groups certainly varied in terms of age, class, education, ethnicity, gender and religious affiliation. However, the purpose of this recruitment process was not to produce a representative sample of Bristol residents, but to access a diversity of viewpoints and opinions. Each focus group was recruited by a local 'host', through personal contacts of the research team or through local services such as community centres and libraries. Hosts were paid a fee for recruiting six local participants and providing a venue for the discussion group. Because of this approach to recruitment most of the focus groups tended to consist of pre-existing groups, consisting of clusters who knew at least some of the other participants through living, socializing or working together. This feature means that these focus groups provide access to the ways in which people talk in the sorts of group settings in which they actually operate (see Kitzinger 1994: 105). This type of focus group research is therefore an effective method for accessing what Shotter (1985) calls 'knowledge of a third kind' – not knowing-that or knowing-how, but

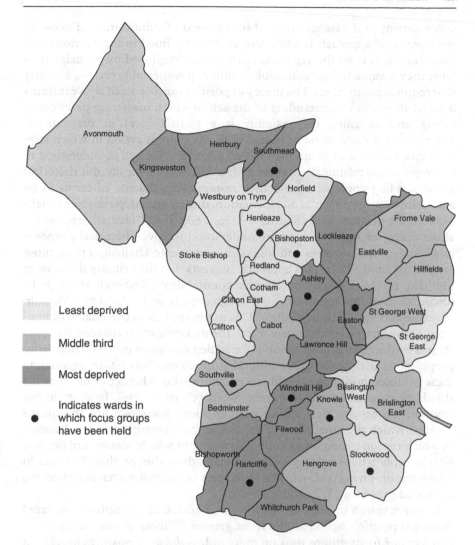

Figure 5.1   Deprivation by ward, Bristol (*Source*: Office of the Deputy Prime Minister, Index of Multiple Deprivation 2004)

'knowledge-held-in-common-with-others', a kind of knowledge 'that one has *from within* a situation, a group, social institution' – 'knowing from' (Shotter 1993a: 19). The names used in this chapter are pseudonyms. Data analysis focused on the identification of recurrent sequences and patterns of talk, rather than focusing on inferring the intentions of speakers or explaining talk by reference to their social roles (see Wetherell *et al*. 2001a).

The objective of presenting our analysis of this talk-data is to identify the meanings attributed to ethical consumption by individuals with no strong commitment to the issues involved in this field of campaigning. Following the 'synthetic' approach to discourse analysis outlined in the previous section, our focus is on the argumentative threads deployed by participants – what they assume to be a defensible position, a respectable reason, a matter that requires justification. The theory of positioning discussed above informs a social-theoretic understanding of the self in which mastering practices of taking and assigning responsibility is a central facet of personhood. Personhood is understood to inhere in forms of joint action in which subjects give accounts of themselves to one another – the self is understood to be 'responsively relational' (Shotter 1993b). Methodologically, this rhetorical understanding approaches discourse responsively, that is, 'in terms of an answering response such as an affirmation, disagreement, puzzlement, elaboration, application, etc.' (Shotter 1993a: 180). Even 'factual' expressions about reality or states of mind are understood to have 'rhetorical purposes in *accounting* for ourselves and others in response to challenges from those around us' (ibid.: 182). This approach suggests that the primary dynamic in the 'take-up' of positions in talk is accountability (Wetherell 1998: 394); discourse from this action-oriented perspective is understood as a 'domain of accountability' (Edwards 2006). This in turn points to two related dimensions of the analysis of discourse. First, the dynamic of positioning is in part shaped by the generation of 'trouble' for settled narratives of self by normative propositions that challenge existing identifications. Second, in negotiating these troubled positions, individuals are addressing whether or not maxims should be considered rules for them (Wetherell 1998: 394). In short, in our analysis, we assume that ethical consumption discourses are the means through which the circulation of 'troubling' positions generates a dynamic in which reason-giving breaks out, a dynamic in which reasons are not just simple explanations of why such-and-such does this or that, but one in which reason-giving is closely tied to the task of justifying what have become contested practices (Tilly 2006: 52–56).

In our research project, the very topic of ethical consumption generated 'troubled positionings' in all the focus groups. Without prompting, participants in our focus groups took up more or less *defensive* postures in relation to this very general topic. In a sense then, the substantive topic of the research projected ahead of us as researchers a 'vertical positioning' of our research participants as presumptively *irresponsible* consumers. In turn, we can analyse these focus groups as modelling the process whereby people jointly consider the extent to which certain maxims about consumption do and should hold for them, by taking their everyday practices as objects of reflection. At the most general level, this indicates that there is a specific grammar guiding talk provoked by reflecting on the ethics of consumption. Our informants interpret 'position calls' by placing them under certain descriptions of reality. To better understand this process, we make use of the concept of 'factual versions'. From a discourse-analytical perspective, the descriptive aspects of utterances are always articulated in a performative register of one sort of other. This implies analysing how factual reports, or claims of fact, are deployed in talk-in-interaction – not least by attending to the process of 'managing negative evaluations via factual versions' (Wetherell 1998: 393). If discourse is understood to always have an action orientation, and if in talk people 'display what they know – their practical reasoning skills and competencies' (ibid.: 391), then this suggests that descriptions in talk-in-interaction are related to participants' concerns (Wetherell and Maybin 1996: 244) – to what they care about and what matters to them. This means that analytical attention should be addressed to investigating the ways in which 'factual versions' – reports, descriptions and representations – are used to perform various actions in discourse (Potter *et al.* 2002: 386–387). Factual versions are an important mechanism for 'doing attribution'; they are deployed rhetorically to support or challenge different normative propositions (ibid.: 389). In short, from this social-theoretic and methodological perspective, there is an inferential relationship between the ways in which speakers use factual versions to assign responsibility to other actors and events, and the concerns that those speakers have with their own accountability (ibid.: 395–396).

In this section, we identify three recurrent 'versions' that arise in focus group discussions about ethical consumption. We analyse these versions as the descriptions under which our participants make sense of and assess the normative claims publicly circulating around everyday consumption in the UK in the mid-2000s. The three 'versions' which we identify in the focus group discussions each revolve around everyday, problem-solving activities: first, there is a version in which everyday consumption is understood as primarily and legitimately determined by calculations of monetary cost; second, there is a version in which consumption activities are placed within the temporal routines of everyday life; third, there is a version in which consumption is placed across a division between hard work and good fun. In each case, these versions articulate claims about where individuals feel they have scope to exercise agency as 'consumers'.

'I have to go for the cheapest . . .'

The first of these versions is indicative of the attribution in which ethical consumption is understood to ask individuals to spend more money on 'ethical' products. This attribution recurs in different ways in focus group discussions, and around a variety of different products, but the 'natural' topic of conversation around these issues tended to settle on food:

RICHARD: I don't necessarily go and get organic veg. I go for the cheapest.
ROBERTA: Yes, so do I go for the cheapest. Because I'd love to go for organic food but I just can't afford it. So I have no choice. I have to go for the cheapest . . .
FACILITATOR: What about you Rhyannon?

RHYANNON: I don't because of the price.

ROBERTA: It's too much that kind of jump, I think, because I do look around Tesco a lot, thinking I should buy organic and, say, fair trade, and I just look at the price and it's just, like, it's ridiculous.[4]

In simple analytical terms, in this exchange Richard positions other participants around the theme of price by saying he goes 'for the cheapest'. Roberta confirms the same principle. The facilitator then locates Rhyannon in terms of this same position, and she likewise confirms its validity. This sequence of talk turns on the relationship between an 'ethical' ought, focused on organic and fair trade food products, towards which participants express an attraction, and a practical 'can', expressed in terms of price, which prevents them from always buying 'ethical'. In the simplest terms, this sequence attributes to ethical consumption the demand that individuals pay more for the ethical choice. In this sequence, the attribution of higher cost is presented as rendering the ethical demand impractical, without any further justification being felt necessary.

A more 'troubled' sequence around this same topic of price revolved around fair trade chocolate:

FACILITATOR: What does everyone think about the idea of paying slightly more for a chocolate bar so people working on the other side of the world get an extra little bit of money when they're doing the farming of the cocoa beans or whatever?

ABIGAIL: It probably wouldn't make a difference. I would still go out and buy the cheapest one. I'll be honest. It wouldn't affect me. It's the money again. I'm so shallow!

[LAUGHTER]

PAULA: But it's true. If you've got a budget to live on, you can't be thinking about people on the other side of the world and then the kids go without because of them, because you want a chocolate bar that costs double your money.[5]

Here, the facilitator's question serves as a 'vertical' positioning of the research subjects around the choice between cost and 'ethics'. Abigail responds by admitting that the 'ethical' aspect would not affect her. But her final, light-hearted remark about 'shallowness' indicates that admitting this quite so honestly might, in this company, transgress a moral norm. Paula takes up the point, confirming the validity of Abigail's reservation by explaining that the realities of household budgeting and family responsibilities render the ethical demand redundant.

Here then, in both these sequences of talk, the 'versioning' of ethical consumption as more costly is straightforwardly presented as limiting the normative force of the demands placed on people as consumers. However, it is not only cost that counts against demands to purchase ethical products.

The women in the following sequence of talk, all young mothers resident in the Knowle ward of Bristol, initially assent to the facilitator's positioning of them as thrifty shoppers:

FACILITATOR 1:   So would you say that the thing that's most important for you when you go round picking things off the shelf is that you're looking at price?

JAN: Yeah, you got to do that.

MICHELE: Yeah, definitely.

SHERRIE: First thing is the price, definitely. . .

FACILITATOR 2:   Name some things where price might not come first.

JAN: What your bloke moans about.

MAGGIE: I do like my Heinz baked beans . . .

JAN: Cheap cereals. It's got to be proper Kellogg's cereals.

SHERRIE: My kids notice the difference straight away.[6]

The participants agree that price is a primary concern for them all, but when further prompted, they introduce a series of additional concerns that also shape their shopping choices. Among the women participants in our focus groups in particular, the attribution of unrealistic costs to ethical consumption was routinely articulated through descriptions of the everyday dilemmas of family life. Favourite brands are mentioned, and in the same breath the importance of brands is located within family relations with partners and children. We see this movement in the following sequence of talk from a focus group consisting of young working-class mothers from Hartcliffe, one of the most deprived wards in Bristol:

JACQUI: I don't look at prices. I just go to the till. I never look at prices, never.

FACILITATOR: So how do you know which brand to buy? Say you're buying washing-up liquid?

JACQUI: I'd always buy Fairy. Just what you're used to, I suppose. If I did go in there and I usually buy Fairy and say there was Asda three for a pound . . . But you just get used to the same things.

MEGAN: I couldn't have cheap shopping. It's got to have the proper name to it.

JACQUI: You just get used to buying the same things, don't you?

MEGAN: That's right.

JACQUI: You get your Andrex and . . . You just get used to it.

FACILITATOR: So is Fairy something that your mum used to use?

JACQUI: Yeah, probably. I used to go shopping with my mum quite a lot when I lived at home, so yeah. And Ariel washing powder.

MELODY: With washing powder, I'm always changing. Whatever's on special offer.

JACQUI: I do sometimes, yeah.

MEGAN: I think you get what you pay for anyway. If you want proper stuff, you got to have a proper name.

MELODY: But in Kwik Save or Iceland, you have a special offer on the washing powder, always. They've got Persil now.

MEGAN: I don't use a cheap washing powder.

MELODY: No, I couldn't

JACQUI: Or cereals, or washing-up liquid or soap or shampoo or HP.

MEGAN: Everything's got to have a proper name on it.

FACILITATOR: Just cos the quality's not as good? Is that what you're talking
about?

MELODY: I've bought these things before and, like, you get three-quarters of
juice.

JACQUI: Yeah but for the kids I suppose it's all right isn't it? Kids don't know
any different, do they?

MEGAN: Yeah that's right, yeah.

JACQUI: When they were little, it didn't make any difference if you bought a
9p tin of beans. But now they're getting older.

MEGAN: It does in my eyes.

JACQUI: Yeah but they're older now aren't they? Depends what you get used to.

MEGAN: Then again, my mum always used to buy proper beans. My dad
would never have anything cheap in the cupboard. It probably goes back to
that. It's probably in your head, isn't it – that that's what you used to have.
And cheap bread. Couldn't eat cheap bread. Got to be Hovis.

JACQUI: Kingsmill.

TRACEY: Depends what you mean by cheap bread, to be honest with you.

MEGAN: You go in some shops and they got their own brands,
20-something-p. . . .

JACQUI: Asda's is 11p a loaf of bread . . .

MEGAN: Yeah, I couldn't eat that.

JACQUI: I buy that to feed the birds![7]

In this sequence, buying particular brands is presented as a matter of the
habitual, routine nature of doing the weekly shop. This also emerges as a
matter of trusting the quality of particular goods. And the matter of quality
is quickly related to the responsibilities that these women feel towards their
children, a responsibility that is presented through the acknowledgement by
Jacqui and Megan of a tension between feeling the need to buy baked beans
of a certain quality and the suggestion that children don't know the difference anyway.

In the extract from the Hartcliffe focus group, the first factual 'version'
through which meaning is attributed to ethical consumption is one in which
questions of price are at the forefront. But price quickly emerges as a means
through which people can start to talk about the range of commitments that
constrain and shape their activities as consumers. The sequence of talk above
illustrates a common feature across these focus groups, which was the extent
to which talking about consumption involved participants in having to talk
about their relationships with family members, friends, neighbours, or work
colleagues. There are three broad modes of referencing this dependence of a

person's own consumption behaviour on the relationships in which their lives, their cares and concerns are embedded. One way of doing so was to narrate a significant 'fateful moment' in a person's biography that stimulated a reconfiguration of consumption activity. For example, having children made a difference to one of the participants:

ROBERT: My girlfriend and I had a couple of kids about 10 months ago, twins. And we buy more organic now cos of them, so I suppose that's changed. Maybe we would have done a bit before but I think now we are just thinking about what they're eating for health reasons.[8]

It is a significant moment in Robert's life, starting a family and becoming a father, that provided the reason, and opportunity, for a change in routine consumption habits. A second, more ordinary, form of marking the 'relational' qualities of consumption activity was to talk about how much of one's shopping was done with friends. For example, for some of the women in our focus groups, this was a matter of the time available during the week when their children were at school, or on Saturday's, when husbands and boyfriends were out watching the local football team. And, third, focus group participants also talked about how they learnt about the 'ethics' of different products not from formal information campaigns, but through social networks: from friends, from church groups, or from what their children told them about what they had learnt at school.

These recurrent themes in the talk-data indicate a widespread attribution to ethical consumption of potentially onerous demands that run counter to individuals' other commitments. While the price of ethical products is often invoked as a problem, it quickly becomes evident in the course of talk-in-interaction that 'price' opens out into a series of domestic and personal commitments that shape people's shopping activity in particular. These focus group discussions therefore confirm the idea that everyday consumption is far from a matter of 'consumer choice' amongst rational utility maximizers (see Miller 1998). People talk about their consumption as a function of their affective attachments to others. Nor, it turns out, do people necessarily appreciate being constantly bombarded with information about what is good and bad for them, and what is good and bad about their own choices more widely. Our focus group participants do not respond to information about products as rational choosers. They are just as likely to express exasperation at all the information they are expected to process as consumers:

CLAIRE: There's something different each week. 'Don't eat chicken' this week because this, this, and this.[9]

The exasperation is often articulated in a register that tightly delineates the scope of 'choice' that people should be expected to exercise. When people are asked to justify their consumption behaviour, they quickly turn to justifying their commitments and relationships. And in so doing, they rarely if ever talk about being a 'consumer'; they talk instead about being a parent, a friend, a spouse, or a citizen, an employee, or a professional.

'At the end of the day ...'

The factual version in which the price of ethical products renders the moral demands of ethical consumption superfluous is linked to the second recurrent factual version which emerges from our focus groups. In this second version, ethical consumption is described as requiring individuals to spend a lot of time and energy in making 'ethical' choices over all sorts of consumption activities. Here, then, the attribution given to ethical consumption is that it requires people to adopt an overly exacting model of how to go about the tasks of everyday social reproduction. This version is explicitly articulated by a number of our focus group participants:

RICHARD: At the end of the day, people have to get to work and they've got their lives to live. They've got to pick up kids from school and live in this super-fast lifestyle which needs to get to places really quickly and, you know, get loads of things.[10]

NICK: Why do people go to a fish and chip shop when you can cook in your own house? People are too busy nowadays. They find that they've finished work, they've had a hard day, and think 'Oh, I've got to go and cook now'. Whereas you can just ring up for a pizza or get a few chips down the road and everything is ready.[11]

PAUL: I would love to say that I hate Tesco's and I never go there. But, at the end of the day, when you've only got ten minutes, and Stellas are £10 for 20, you've got to go. It's just far too convenient, especially when you're busy, when you have a busy lifestyle.[12]

In each of these statements, the busy-ness of everyday living is presented as a fact that militates against expending too much time sorting through the ethics of consumption. None of these participants is expressing any great sense of personal fault about this state of affairs; a sense of regret perhaps, but they are each describing as a factual state of affairs the felt pressures guiding the consumption habits of themselves and others.

In the second factual version, in which ethical consumption is described in relation to the temporal routines of everyday life, themes of busy-ness and convenience are presented as practical limits to responding to the demands to monitor the 'ethical' qualities of everyday consumption activity. One aspect of this factual version is that it often appears alongside expressions of ambivalence towards the value of 'choice'. The ambivalence that people express about the value of choice is neatly illustrated by discussions about the advantages of vegetable box schemes (see Clarke, Cloke, *et al.* 2008). These can be a convenient way of getting vegetables and fruit delivered to the doorstep and being 'ethical' in an organic way at the same time. Some people don't like the lack of choice implied by these schemes:

CAROLE: I knew someone who has one of those boxes that you're referring to, and she's very pleased with it.

STEPHANIE: I know somebody and she's thinking of cancelling it because there's only two of them and they've no control over what goes in it so they get rather a lot of what they've got a lot of and sometimes it's not always what you want.

JANET: They can't specify what they want, then?

STEPHANIE: No you just get a selection.

DAVID: Of what's available, yeah.

STEPHANIE: So they're thinking of cancelling it.

CAROLE: You can choose what you want from ours.[13]

Others appreciate the lack of choice involved in these schemes, because they claim that it adds both a kind of surprise and a kind of obligation to their everyday cooking activities:

MICHAEL: There are veg boxes, organic veg boxes you can get.

RACHEL: That's true. Yeah, that's true: you can just go pick it up on a Thursday night or whatever.

NIGEL: Which one do you get?

SIMON: Green Wheel.

RACHEL: Any good or mouldy?

SIMON: No, it's good. It's £10 for fruit and veg for two for a week and there's always potatoes, onions, carrots and then odd greens and things and enough fruit to last.

RACHEL: I like the way they just arrive and you don't have to have that thought about shall I buy that or not?

SIMON: It forces you to eat more fruit and vegetables.

RACHEL: Exactly . . .

SIMON: Because you think I can't chuck out . . .

RACHEL: Not bloody broccoli again!

JOHN: So you don't have a choice what you get; it's just thrown in?

SIMON: Yeah, but there's always potatoes and onions and staple things. That's part of the joy: it's interesting new things arrive.[14]

Here, signing up for the weekly delivery of organic vegetables is presented as a way of forcing people to eat more healthily at the same time as making cooking a little more interesting. In this respect, this exchange nicely connects the value of vigorously exercising consumer choice over all aspects of everyday consumption to questions of pleasure and fun. The first two factual versions we have discussed – descriptions of ethical consumption in terms of price and in terms of the time – relate to a third factual version in which ethical consumption is attributed an explicitly puritanical meaning which runs against the ordinarily 'hedonistic' requirements of the good life.

'If you knew everything that was going on . . .'

In the third factual version, the idea that it is a good thing to know the 'biographies' of the goods one consumes in everyday life is treated with considerable scepticism:

> JAN: I don't know half of what's going on. If you knew everything that was going on through all these different places you wouldn't eat.
> MICHELE: If you knew all these things, everything that was going into these different things, you'd have a nervous breakdown, wouldn't you?
> GILL: You'd starve to death, wouldn't you?[15]

There is a sense expressed in this sequence that the demands made on ordinary people and their everyday consumption habits can be too overwhelming. These participants are not making excuses here: far from it, they are presenting the demands of ethical consumption as unreasonable. In so doing, they are delineating the limits of what they consider should be required of them. They express these limits by presenting as a matter of fact the view that carrying on a normal life is not consistent with monitoring every single aspect of one's consumption for its 'ethical' content.

The three factual versions we have identified in this section have some of the characteristics of 'interpretative repertoires', in so far as they contain common tropes that recur across the focus group discussions with Bristol residents. But we prefer to emphasize their qualities as 'factual versions' because each one is deployed in ways in which allow participants in interactive talk to adopt defensible positions on 'troubling' topics that are implicitly and explicitly acknowledged by their peers. In these focus groups, the very topic of 'ethical consumption', which participants turned up to talk about, immediately prompted more or less defensive positionings. These factual versions provided the relatively uncontested vocabularies through which participants delineated the scope of what should be considered an 'ethical' demand on them given the practical constraints of money, time and happiness. Each of these three factual versions also serves as a way of articulating an understanding of consumption as a function of the activities of everyday life – looking after children, commuting to and from work, leading an ordinarily happy life. In short, what is being 'versioned' here is people's practices, in relation to which any demand to change consumption activities is in turn thoughtfully assessed. These versions are not only about what people do, they explain what people do as an aspect of what they can be reasonably be expected to do.

We want to emphasize that the factual versions that circulated in our focus groups are not simply focused on what people are *not* willing to do. They are also the medium through which people articulate how any form of 'ethical' consumption in which they do happen to participate is determined by what they *can* do. Participants in the focus group held in Bishopston, a hub of alternative retailing in Bristol, expressed the sense that the sort of 'ethical' consumption they participated in was a matter of doing things with fitted with existing concerns, energies, and interests. Nobumi articulated this quite clearly:

> NOBUMI: I buy or I do ethical shopping but for mainly environmental reasons, not much of fair trade. I mean fair trade I sometimes buy because probably it's a little bit cheaper on that day or something, or, you know, if I like a certain brand I might buy, but normally it's for environmental reasons. Recently I discovered that shop across the road, Scoopaway, sells cereals, so I've stopped buying boxed cereals now because I really hate throwing all the boxes and plastics and everything so I just go there and put some in my bag and then just put it in my container at home, so then I reduce the cardboard box and plastic bag. And that's my reason and I think it's good that people have different reasons to do ethical shopping and it probably doesn't matter because at the end actually you're contributing to the entire nation's attitude. I think it's really important, I don't know whether there's any wrong reason to do it![16]

Nobumi affirms the idea there are different sorts of 'ethical' commitment, and openly acknowledges her own commitment to environmental issues rather than fair trade. The sense in this comment that there are a variety of reasons which motivate the adoption of ethical consumption is supported by another participant in this focus group discussion:

> JOHN: Ultimately everyone thinks about themselves I think and it saves you money, affects your local environment, you go for it, but it's very hard to see the effects of fair trade; you don't know, it might be improved, it might not be, it's hard to say, if it helps me, if it was demonstrated, what was going on, then it would make me feel better but you could actually even argue if it makes you feel better socially it's a selfish act as well. They always are selfish acts really, ultimately, because you're doing something to make you feel better, even you said didn't you that you felt better for it. That is a kind of selfish act isn't it? You make yourself feel happy about an issue. So I think those are the best ones.[17]

John is thinking out loud here about the degree to which acting 'ethically' should require him to be self-less, and he ends up affirming the idea that the 'best reasons' for acting ethically are the ones which 'make you feel better'. It is notable that he constructs fair trade as less compelling not just because the effects of this type of consumption are hard to see, but also that in so doing he presents fair trade as a model of ethical consumption which requires the adoption of a more other-regarding perspective than he feels is either plausible or preferable. Nobumi also contrasts fair trade, to which she is not strongly committed, to the adoption of consumption activity that is focused on the tangible reduction of the amount of stuff she throws out. For both participants, then, fair trade serves as a reference point around which they account for the fact that the 'ethical' activities which they do engage in are directed by the *partial* commitments which motivate them.

Nobumi and John are both accounting for the sorts of ethical consumption they *do* engage in, rather than for why they do not. But in doing so they articulate the same sense that changing one's consumption activities is a matter of doing what is practical that we found amongst participants who are less engaged and more defensive. And in acknowledging this, these participants give voice to a more modest understanding of their own 'ethical' reasons than is often associated with committed 'ethical consumers'. They disavow strongly altruistic accounts of their own motivations, in favour of accounts in which they do what they feel able to do for reasons which they justifiably feel include their own well-being.

Daniel Miller (1998) has argued that what he calls 'moral shopping' is embedded in relationships with loved ones, friends and colleagues, and that this is often experienced as being opposed to the demands of 'ethical shopping' in the name of principles of equity, fairness and justice for anonymous and distant strangers. In this section, we have seen this sort of distinction expressed by focus group participants both as a reason not to engage in ethical consumption, but also as a reason to account for the sorts of activities more engaged individuals do feel committed to. As we saw above, fair trade is often invoked as the model of a strongly principled form of ethical consumption against which other forms of activity are contextualized as being more in tune with practical and self-regarding concerns which are presented as quite appropriate reasons for doing what one can. In line with Miller's account, Thompson and Coskuner-Balli (2007) argue that ethical consumption initiatives work to 'enchant' what might otherwise be considered inconveniences, restrictions and costs of adopting new consumption habits by reconfiguring these as virtues that can be articulated with the concerns of everyday 'moral shopping'. Our own analysis in this section indicates that efforts to enchant 'ethical consumption' in this way have to negotiate the constraints expressed in these factual versions, constraints which delineate the space in which respondents felt little compulsion to offer strong justifications for not considering 'ethical' demands as binding on them. Rather than interpreting this type of talk as a form of buck-passing, we have presented it here as a form of first order practical reasoning in which participants assess ethical 'oughts' in terms of pragmatic modalities of 'ability' and 'competence' (see Törrönen 2001). In the next section, we further develop our analysis by considering a second order of practical reasoning evident in the focus group discussions. In this strand of reasoning, participants begin to explicitly engage with and sometimes contest the normative content of the demands addressed to them through publicly circulated ethical consumption discourse.

5.4   Dilemmas of Responsibility

The previous section showed participants in focus groups engaging primarily in the *practical* side of practical reasoning about the political and ethical aspects of their everyday consumption activities. This section shows how they engage in practical *reasoning* which explicitly addresses the authority of the claims made on their activities to count as reasoned in the first place. We consider how people engage with the position calls of ethical consumption discourses to take responsibility for global issues by reasoning about the dilemmas that these positions place them in. Here we focus on the sequences of talk-in-interaction in focus groups in which participants explicitly question, argue and elaborate on the extent to which various maxims are binding on them. In line with the rhetorical approach outlined at outset of this chapter, we analyse utterances as responses to positionings, responses which affirm, question, clarify or apply. As we discussed in Chapter 2, if we conceptualize ethical consumption campaigning in terms of the concept of ethical problematization, then we should understand the demands circulated in public discourses of global responsibility as addressing various dilemmas to the public – as we saw in Chapters 3 and 4, the rationalities of these campaigns increasingly presuppose that ordinary citizens are argumentative subjects.

The elaboration in the previous section of the factual versions deployed by our informants indicates that ethical consumption discourses position people as potentially responsible for all sorts of global issues. It shows too the ways in which people delineate the scope of their activities liable to such demands by reference to practical considerations. In the course of these focus groups, discussion often shifted into explicitly normative registers. This shift took place in two entangled but analytically distinct directions. First, participants would begin to elaborate on the *complexity* of the moral demands being pressed upon them. And, second, they would explicitly question the ascription of responsibility to them as *individuals* made by public discourses of global responsibility. In both respects, what we can observe is participants shifting registers to evaluate the 'ethical' demands made in terms of a 'deontic' modality by considering whether these demands *should be* obligatory for them, not only whether they *could be* (see Törrönen 2001: 321).

In the focus group held in Easton, one of the young male participants bought up the example of campaigns against Shell. This is an instance of horizontal positioning, in which participants introduce an example from their own experience around which to develop their own thoughts and generate further discussion with their interlocutors. In this case, Nick had learned about the Shell issue when he was canvassed by campaigners at a local petrol station:

> NICK: I am a person like this, I only act if I see something done about it, like, see me if I'm willing to find out how much information is being done about them animals getting badly treated. Do you know what I mean? Like, if I'm willing to go and find out about it and find out if I know if the changes are being made or I can help with the changes, well I'd stop it, but if I can't and I'm just seeing a leaflet and my car has run out of petrol and I'm outside a petrol station then I'm going to use it and buy a packet of quick tricks as well.
>
> ALEX: See they give you leaflets and tell you what they're doing but because I actually had one of those leaflets but not one place on that leaflet did it tell you what you could about it. So we know this is happening okay, thanks, yeah nice, what am I supposed to do? Know what I mean? I haven't got no petrol in my car.
>
> NICK: You even see people standing outside Shell campaigning, not to buy your petrol from there. But at the end of the day it's not like it's happening there, no, you have to buy petrol; if you've got a car, it's a necessity. You have to buy petrol. At the end of the day, even though the people might be standing outside Shell or giving out leaflets, BP are probably doing the same thing, or other petrol stations might be doing the same thing but you don't know. And if you're going to stop everything you're not going to get no petrol.
>
> ALEX: You're going to lose out. When you don't make it to where you want to get to, you lose out. And if you don't make it to wherever you want to, you're the one who's going to get in trouble.
>
> NICK: There's controversy about everything. Everything.[18]

A significant shift takes place in the course of this exchange. Nick begins by defending what he currently does with reference to the argument that claims that he should change his behaviour must be ones that can be acted upon in a practical way. When Nick introduces the example of people campaigning outside of the Shell service station, he and Alex continue to articulate the line of reasoning according to which they have to continue to go about their everyday activities – they need to buy petrol from somewhere. As they express this view, though, they begin to articulate the sense that perhaps there is not a straightforwardly 'right' thing to do in this case after all – that BP might be just as bad as Shell, and that they cannot know for sure. Nick then concludes that everything is 'controversial'.

This sequence of talk therefore revolves around the intuition that no choice can be entirely 'ethical', since there is no clear agreement on what the 'ethical' thing to do is in any particular case. In the focus group discussions, the sense that choosing the 'ethical' option is far from straightforward is often initially expressed as being a practical matter. This is what Nick and Alex are doing in the sequence of talk cited above, as they agree with each other that they cannot really be expected to boycott petrol stations because they still need to get around in their cars. But this practical, often defensive, register is frequently supplanted by a shift to a more explicit consideration of whether what is publicly presented as 'ethical' is actually a simple issue. As we have already indicated, participants in focus groups introduce a wide range of their own examples of consumption activities as bearing on 'ethical' issues. This is often done to exemplify and elaborate on the complexity of what being 'ethical' actually means. Examples introduced 'horizontally', by participants themselves, are often jointly elaborated to develop a shared understanding that there is a series of tensions between contradictory ethical demands made upon people as 'consumers':

ROBERT: There's so much information about everything these days.

PAUL: You just to have to make your decisions and do what'll fit into your lifestyle. If everything that was greener was no more effort and cost no more then everyone would do it.

ROBERT: But people do have to give a bit to be greener, definitely . . .

PAUL: But if you were to do it with everything you just wouldn't have the money . . .

GAVIN: Yeah, people just think that if I have to put any more effort in then I'm not going to do it, which is wrong. The issue is that when is it greener. Those underlying problems you come across. Once you start looking into it, it really does become a minefield.

JACQUI: I heard recently about a fair trade coffee and apparently in some places because fair trade coffee is moving more quickly from some areas, gangsters are coming in and taking over these coffee plantations . . .

ROBERT: There's always another way of looking at it, it's true.

ANDY: We have so much information; it used to be the great thing, everyone must have a choice of whatever they do. But people nowadays, in my experience, they don't really. They just want some free things to choose from. And personally I find it quite useful for even a vegetarian to go to a restaurant and there's only two choices. That's one less thing I've got to think about.[19]

This sequence of talk about what counts as 'green' starts out focusing on issues of practicality, such as having enough information or having the money. But Gavin introduces the idea that actually being 'greener' is not so simple, that it might be a 'minefield'. Jacqui then introduces her anecdote about fair trade coffee and gangsters as a way of affirming this intuition that the choices that are coded as 'ethical' might look different from another perspective.

In these focus groups, there is a characteristic way in which 'information' about the ethics of consumption is processed in talk-in-interaction. A piece of information – often referenced as something read about in newspapers, seen on television, heard on the radio or picked up from friends – is introduced into the discussion as an example around which participants jointly consider the moral dilemmas that it throws up. When participants introduce their own examples in this way, it is often to reflect on just how difficult it is not only to *do* the right thing, but also how difficult it is to *know* what the right thing to do actually is. Buying products made in far-away sweatshops is the most commonly introduced example around which this dilemma is framed:

KAREN: I worked for a company that switched from products manufactured 20 miles down the road to having products manufactured in China, because they were cheaper, and they argue that they're cheaper, and then businessmen that went over to China, stayed over, exorbitant trips, and add all that together and it's apples and oranges in the end. But I saw photos of how the Chinese workers lived and, well, I was told their circumstances and the pay they got was better than the surrounding environment and that's one of the things that I think about, well, the company I work for: did they do a good thing or a bad thing, what's better? Because they're getting tuppence more than somebody else is that good – just because it's better than crap, is it good? I struggled with that.

SIMON: It's a very difficult question.

KAREN: It is difficult.

JOHN: If you buy these shoe things in a sweatshop, you say they're exploited but by not buying those shoes, that person loses a job, so they're exploited at one point but at least they have a job even if they're exploited.

ARUN: That's the argument the big corporations use.

KAREN: Somebody's going to exploit, it might as well be us.

JOHN: I know, but it's partially valid isn't it? Okay they're exploiting them but they've got a job now, so the question is: is your act of kindness making it worse than it was in the first place?[20]

The participants in this discussion are neither excusing nor justifying their own conduct. They are using the example as a focal point for a relatively abstract reflection on how performing one action coded as 'ethical' might have consequences which, from another perspective, might be considered 'unethical' – in this case, wrestling with the dilemma of whether buying shoes made in 'sweatshops' is complicit with exploitation or to support people by maintaining them in employment. In the case both of Karen's personal anecdote about her former employer and John's reference to the generic 'sweatshop', examples are being deployed here to focus attention on the 'difficulties' of acting ethically as a consumer, not as straightforwardly factual pieces of 'information' from which an obvious path of action automatically follows.

The sequences of talk analysed so far in this section involve participants rhetorically processing information about the ethics of consumption: that is, by considering the types of dilemmas that information raises for them about the best course of action to take. In these cases, participants raise the possibility that consumption activities involve a degree of complexity which calls into question the possibility of a simple evaluation of what making the 'ethical' choice is. Sometimes this involves questioning whether consumer choice can ever really be entirely 'ethical'. Sometimes it involves elaborating on the intuition that choices coded as 'ethical' might turn out to be less 'ethical' than they appear. And sometimes it involves reflecting on the possibility that even 'unethical' activities might be less blameworthy than the moralistic register of ethical consumption discourses often suggest.

We have seen in this section that focus group participants shift to explicitly normative registers by introducing the idea that the moral demands made upon them in everyday consumption activities are complex. This sense of the complexity of the ethics of consumption is a common theme of the focus group discussions conducted in Bristol. It is a theme that tends to elicit agreement amongst participants. But this expression of the complex *form* of ethical decision-making is sometimes also related to the emergence of a much more explicitly contentious form of talk, in which the *content* of the moral imperatives of ethical consumption discourse becomes a topic of reflection. This shift is most frequently provoked when the topic of 'fairness' is raised in the context of discussions of fair trade products. The following sequence of talk is from the focus group held in Hartcliffe, involving young working-class women:

FACILITATOR: There's another group of people who may choose not to buy certain things because of the way the workers who produce those things in the factory get treated by their bosses. So say, for example, there was a story a while ago, I don't know if you guys heard about this, that Nike was using child labour to make their trainers and their footballs and stuff like that, over in Malaysia and Indonesia. I don't know if any of you guys heard that story.

JACQUI: I sort of read about it.

MEGAN: It was in the paper wasn't it?

FACILITATOR: Yeah. Would that bother you guys or would you consider buying something different if you heard a story like that? Do you care?

MELODY: Well, you do care but you don't think about it. If you like a pair of trainers you're not going to think . . . If you like them you'll try them on . . .

FACILITATOR: So it's just something you don't want to think about?

JACQUI: Well, you do care about it but. . .

MELODY: It's cos it's not happening here, is it?

JACQUI: It's cos you don't see it.

MELODY: It's not something you know.

FACILITATOR: So it would be different if the factory was down the road and some of the kids that you knew who were knocking around Hartcliffe were working in the factory?

MEGAN: Yeah. I'd probably feel different about it then.

TRACEY: We'd all buy it then cos the kids would be working! [Laughing]

JACQUI: Get them off the streets!

TRACEY: That's right!

MEGAN: Get them in the factory and make them work!

FACILITATOR: Let me ask the question, is it cos you guys think you don't have any control over this anyway; if you stop buying those trainers it won't make any difference to the kids?

JACQUI: Definitely. It wouldn't make any difference.

MEGAN: There's always someone gonna buy it isn't there? There's always someone willing to buy it.

TRACEY: I don't think it makes that much difference as an individual.

JACQUI: No.

TRACEY: The only difference it would make is if there was a big crowd of people and say, right, we're not going to. But individually, it wouldn't affect anybody at all.

JACQUI: 'Cause they did say at one point in school don't bring any trainers 'cause some children go to school with all these posh £70 trainers and you've got other little kids going down Peacocks and buying them for a fiver. And I think they're worried about it at the school. Wasn't that sometime last year? Something about stopping named stuff? Reebok sweatshirts and stuff? Some kids were getting picked on cos mums and dads couldn't afford to buy their children Nike trainers.

MELODY: That's why you got a uniform.

JACQUI: Yeah, I think that was something to do with it, I'm sure it was. We had to fill in a questionnaire a couple of years ago about that. Now they have. Jessica's got a pair of Reebok trainers but at that time they never . . . they just used to wear the little ones from . . .

MEGAN: You just have what you can afford. If you can afford to dress your kids nice then it shouldn't be a problem.

JACQUI: Yeah.

MEGAN: If you can afford it, they should have it, shouldn't they?

JACQUI: Children can be very cruel. But then to me, why do the children get cruel like that? Who makes them cruel? To me sometimes I think it's the way you've been brought up.

TRACEY: If you're well enough off to buy your children all this expensive stuff and another family they can't afford it, then the kids pick on them because they're, 'Oh, what you wearing that rubbish for?'.

MEGAN: It's not the kid's fault.

MELODY: Our D, everything's got to have a name. Even boxer shorts. They cost me £10 each for them boxer shorts.[21]

We reproduce this long sequence of talk here because it illustrates the interactive dynamics of ordinary reasoning which we are concerned to elaborate on in this chapter. The first part of this discussion, revolving around buying goods produced by child labour, is prompted by the vertical positioning of the facilitator's question. The response is initially focused on what is a quite common line of discourse about not feeling the need to care as much about events going on far away. But as the discussion continues, Jacqui reorients the discussion away from the simple issue of whether she and her friends care about child labour far away. She does so by making a connection between this abstract question to a more concrete, local issue of how this relates to parent's concerns for their children's experience at school. In this move, the generic example of distant sweatshops is bought much closer to home by Jacqui's intervention. And the other women in the group take up this topic in their turn, raising questions about their children's attachment to brands, the affordability of branded clothes compared to a school uniform, and the cruelty of children to one another. These women give voice to the intuition that the relatively abstract demands of 'ethical shopping' and the more caring concerns of 'moral shopping' might actually be related to one another rather than straightforwardly opposed.

The preceding discussion continued with the facilitator asking the participants about buying fair trade products at their local Co-op (the Co-op is the leading retailer of Fairtrade certified products in the UK):

MELODY: I thought fair trade was, like, fair to the customers. When I saw that, I thought well what a load of twaddle. You don't get fair trade at the Co-op, you get just ripped off. It's so dear in there.

MEGAN: The most I'll buy in there is milk and a paper. That's it.

MELODY: They sell the papers there, that's why you have to go in there.[22]

Here, both Melody and Megan express some scepticism about the 'fairness' of shopping at the Co-op, which they present as being expensive and a 'rip-off' – and not therefore 'fair to the customers'. The topic of buying fair trade products at the Co-op recurred later in the same focus group:

FACILITATOR: Let me just go back to this fair trade thing for a minute. You said the problem with the Co-op is it's always five, ten pence more expensive than elsewhere and that's not very fair on the customers. What if that extra five or ten pence went to the workers who were producing the stuff. If you knew that, would it make a difference?

MELODY: No. Because you sell a product because you want your customers to buy it. So why should we have to suffer because of these people who's making the coffee, if you see what I mean? Cos we're suffering at that shop, we're not suffering at other shops. Why can other shops do it? And they can't say, like, it's only a one-off because there's Co-ops everywhere. For donkeys' years. And it's still expensive, even Pioneer, it's not as expensive as that one, but it's more expensive than Asda or Tesco. So really, they're not saving us anything. I think it's wrong . . . We're the customers.

MEGAN: At the end of the day, they know that is all we have to go. There's a lot of people round there, that's all they got. It's their lifeline, isn't it? . . . So they've got to shop there, they know that.[23]

These women, low-income residents of a deprived ward of Bristol, are quite unapologetic in their refusal to affirm the proposition that 'paying a bit extra' is fair, because they do not consider that it is fair *on them*. They explain that the Co-op is the only local shop in the area – the other retailers they refer to, Asda and Tesco, are a car or bus ride away – and they present this as a source of the Co-op's exploitation of them as consumers with limited choice.

The same issue, of whether and why to buy fair trade products, was also addressed in the focus group held in an inner city area:

KURT: I was in a shop in St Werberghs last night and I was looking at the cartoned juices, and shit, fair trade orange juice, I've never seen that before and it made me think. I thought it was just tea, coffee, chocolate but obviously there's a whole multitude of it.

KIM: It's the Co-op that's fair trade isn't it? I wish there was a Co-op round the corner, not Tesco.

RUBY: And they're cheaper anyway Co-op! And they're doing it, Waitrose are doing it now but Waitrose have got the advertising by using all these lovely ethnic minority people and I just think no, you're actually just exploiting people all over again by using them in your adverts. They're advertising themselves as being the same price as Asda and Tesco, that's what they've got the audacity to do, they're advertising themselves as being the same kind of price: it's, like, no you're not! It's terrible, using the people, the actual people as their pulling power; I've seen other fair trade places, cartons of juice, they use the people to advertise the goods and you feel like you're being made to feel guilty and you look at the price and you get two orange juices in Grosvenor Street, ordinary orange juice for a pound, two cartons of orange juice for a pound, and one of those fair trade organic juices costs, like, £1.50 and it's just, like, and they use the people to advertise it, and it's just, like, you get two for a pound. Obviously you go for two for a pound.

GARY: You can't afford it; don't know what you'd go without it or just buy the one you can afford. I don't know. Fair trade to us as well as them. Makers of it as well, it's gone through the channels and ends up on Tesco's shelf for you to buy it but who's it being fair to?

RUBY: If organic food and fair trade food was cheaper ordinary working-class people would buy organic and fair trade because it seems to me I think if you're working class it doesn't matter because you're too poor for it to care anyway if you buy organic or not, just buy with the masses, just eat the poisoned food. If you're middle class, we're gearing this at you but not for the working class because we know you can't afford it but if we could we would.[24]

This sequence of talk reiterates the themes we have already noted. Ruby articulates her doubts about the aesthetics of fair trade advertising. She says that they make her feel like she is being made to feel guilty, and argues that the use of visual images of members of producer communities is itself exploitative. She also introduces an example of fair trade products being more expensive than other products. At this point, Gary interjects by raising the question of whether it is 'fair' to expect people like himself, on relatively low income, to pay more for fair trade products. Ruby expands further on this point, reiterating the attribution to fair trade of being too expensive. Her specific complaint is that she would like to be able to buy more fair trade and organic products, but does not feel able because of having to live on a limited budget. In the course of this discussion, then, Ruby reframes the issues of global responsibility in two ways: she herself introduces the topic of guilt, but does so in the course of voicing doubts about 'being made to feel' like this by the use of particular sorts of images; she then goes on to clearly articulate a reason why she cannot buy fair trade products all the time because of their price. In both cases, the implicit moral demands surrounding fair trade and organic products are being subjected to criticism. The expression of scepticism towards ethical consumption is, then, far from being a preserve of those people looking for excuses for not 'doing their bit'. It is just as often people committed to some type of ethical consumption activity who raise these doubts, or as in the case of Ruby above, people who feel they would like to do more but do not consider themselves to have the resources to do so.

We close this section with two gobbets of talk-data that illustrate the sorts of ordinary reasoning which, we are arguing, people bring to bear on questions of where their responsibility lies for the consequences of consumption activities. In the course of the discussion in the Stockwood focus group, two of the participants, Nancy and David, begin to argue about the ethics of fair trade. The discussion begins with Nancy, telling everyone that she writes letters as part of Oxfam's trade justice campaigns and donates to Christian Aid. This prompted David to raise doubts about the value of giving money to charity 'because you don't know how much is actually getting to the people in need'. This point was then picked up by the facilitator, who asks David directly whether he feels the same way about buying fair trade products:

FACILITATOR: And do you see fair trade in the same way, when you're buying fair trade tea, coffee or chocolate?

DAVID: I think fair trade is a good thing but then if you're bypassing another country you're taking away even though they're only getting a meagre living out of it, you're taking that away from them as well, which is putting . . .

NANCY: Who?

DAVID: . . . the other countries; you're going to other countries that are not giving fair trade and . . .

NANCY: Fair traded goods have bypassed the middle man, not a country; you're bypassing the middle man . . .

DAVID: But . . .

NANCY: . . . to make sure that people who grow the coffee get a fair price and that the majority of what we pay gets to the people that grow the coffee and doesn't stay with the middle man.

DAVID: Right, but if you're bypassing another country because that rule isn't in place, you're withdrawing the demand from that country, which is reducing the amount going to the poor ones, that are getting a very minor pay.

NANCY: No you're not,

DAVID: Yes you are.

NANCY: They're still – their coffee's still being marketed by Nescafé and . . .

DAVID: But you're reducing it; you're reducing it by bypassing it so you're reducing their output.

NANCY: You're not.

DAVID: Yes you are, you're bound to. If you're not buying it, it means . . .

NANCY: Fair traded coffee's being produced, Kenco's being produced; if I buy fair traded coffee, I'm not buying Kenco.

DAVID: So you're reducing that amount.

NANCY: But if you're buying Kenco, you're reducing the fair traded coffee; it's a free market. All I'm saying is I'm buying fair traded coffee because more of the money that I pay for the coffee gets to the person that grows it.

DAVID: But what I'm saying is you're reducing the one where the country is, where they're only getting a small percentage anyhow, and reducing the amount they're getting which means their poverty is going worse and not better.

JOAN: Yes but if the big companies suffer a little bit or find that they're not making so much profit, maybe they'll come in with the fair trade scheme.

DAVID: Maybe, but in the meanwhile the poor people in that country are going through . . .

JOAN: Fair trade's in most countries isn't it?

NANCY: Yeah.[25]

This sequence of talk is notable because it involves a quite explicit *argument* between participants about the ethics of fair trade. While we have already seen that focus group participants are able to articulate disagreements and differences about the topic in front of them, this rarely took the form of a direct person-to-person engagement of this type between Nancy and David here. Nancy, a retired primary school teacher who is actively engaged in ethical consumption, sets out the case for why fair trade is good for producer communities, and is supported by Joan. David counters this position by arguing that fair trade must be reducing demand for products somewhere, so that it might just be making things worse. What becomes clear as this discussion continues is that what is really at stake here is two views about where the responsibility to act should be located. Joan and Nancy saw their own support for fair trade as a way of supporting a scheme that provided competition for 'big companies', and therefore helping to put pressure on them to change their business practices:

JOAN: If these big companies get some competition, they'll have to do something to make their profit, to keep up, so I think if there was more trade, more fair trade business, then I think they would probably have to join that sort of scheme or do something similar otherwise

DAVID: In the meanwhile, it's the poor people who are suffering from that. It would be better if there was more government legislation against that type of company who are exploiting. It would be better for governments to do it, not for us to withdraw.[26]

David does not withdraw from his own sceptical position, but he does make clear that his main concern is that the responsibility for changing global trade systems should lie with 'governments'. This is, then, one example of a common theme of these focus group discussions, in which agreement that the ethics of consumption are complex develops into explicit consideration of whether the attribution of responsibility to participants as individual consumers is valid. In the discussion between Nancy, Joan and David, it is David who is animated by the sense that perhaps it is not; Nancy and Joan, on the other hand, see their own efforts as part of a broader effort to bring about change.

We have already seen examples of the ways in which participants routinely express a sense that they can't be expected to 'do everything' on the grounds of time, money and other practicalities. But sometimes these ordinary concerns about when and where choice is a good thing, and the degree to which 'ethical' considerations can or even should enter into everyday consumer choice, break out into more explicit discussions of the 'politics' of choice and responsibility. On these occasions, participants explicitly raise doubts about whether it is their responsibility to act on all possible 'ethical' consequences of their consumption activities. The Henleaze focus group consisted of women in their twenties and early thirties, from professional middle-income backgrounds, and during their discussion they explicitly reflected on the question of whether they should be considered *responsible* for all the things that, in their view, ethical consumption discourses seemed designed to make them feel *guilty* about:

ALEX: I think part of the reason I go to a fruit and veg shop is I don't really want things that have been in a poly bag or things that have got completely excess packaging; it is, it's ridiculous the amount of packaging you get.

DAWN: We are meant to feel guilty about it, like it's our fault but it's not. I don't choose to buy apples in a poly bag.

ALEX: But it is our fault if we choose to buy apples in a poly bag.

DAWN: It's the people who produce it who are the guilty ones, not us.

ELLEN: At Tesco, I feel that you can't blame Tesco's for knocking down suppliers: most of it's their customers that are demanding the prices are the same.

CLAIRE: Sainsbury's where they are doing the beans, they used to sell them all, whatever. You put them in a bag. So you pick out what you want, but they have completely got rid of the ones they used to sell loose and now they only give you the packaged option. If you haven't got that choice . . .

ALEX: If you go to the greengrocer you get, you take your own poly bag, which I don't, but you get a poly bag and you say I want those potatoes, she puts them in the bottom, and I want those apples, they are all just chucked in the bag, fine. That isn't anything extra at all.[27]

While Alex articulates the position in which she, as a consumer, can make a difference, Dawn is more sceptical, saying that 'we are meant to feel guilty', and going on to say that 'it is more the system, we are part of the system' and that 'there is a pressure put on us to feel responsible for everything that has happened'. The women go on to talk about the pressures and influences that shape consumer demands, mentioning television, adverts and fashion. In this way, they place the role of consumer 'choice' in a broader context. This style of responsibility-talk should not be interpreted in terms of participants trying to displace their own obligations. These women all agreed that they should try to 'be aware of what is happening and try as much as possible'. But, as in other focus groups, the primary tone of the discussion, evident in the refrain about 'being made to feel guilty', which we have already seen in other focus groups, expresses unease at the idea that so many issues are being presented to them as matters of individual responsibility.

The idea that responsibility for various 'global' issues might not actually lie with the consumer is, we have seen, a topic that emerges throughout the focus groups: we have seen David in Stockwood articulate it in one way, Dawn in Henleaze in another, and Ruby in Ashley in yet another. In the course of these sceptical conversations, the idea that 'consumer choice' can or should be a mechanism for bringing about changes to environmental issues, human rights abuses, or global trade injustice is being subjected to various 'tests'. Our participants introduce their own ethical criteria to consider whether various demands are obligations on them – criteria of practicality, coherence and equity, among others. And, sometimes, the unease about 'being made to feel guilty' breaks out into explicit arguments that responsibility for the range of issues problematized under the topic of ethical consumption does not primarily lie with individuals as consumers at all. There is a consistent pattern to the focus group discussions of ethical responsibility and everyday consumption. Discussions start off by focusing on the practical limits of people's capacity to act on the 'ethical' demands being addressed to them as 'consumers' – participants talk about whether they *can* act 'responsibly'. But as discussions proceed, this focus shifts to an explicit consideration of whether all these matters are the responsibility of individuals – participants begin to reflect on whether these things *should* be matters of personalized responsibility at all.

In the preceding section and this one, we have identified two dimensions of responsibility-talk, understood as a form of ordinary practical reasoning. In focus group discussions, we find people delineating the scope of their daily activities which they feel able and willing to subject to certain forms of ongoing moral reflexivity. Sometimes, they frame the moralized address surrounding consumption by adopting various rhetorical modes of irony, denial, regret, excuse-making or justification. This leaves the content of the moral demands unchallenged. But sometimes we can catch them directly contesting the idea that consumption habits should be regarded as bearing these sorts of moral burdens in the way that is increasingly expected. We could easily interpret this style of talk as a means by which people displace and deny responsibilities that they should, ideally, be willing to acknowledge. That is what lots of policy and academic research is inclined to do, as we discussed in section 5.2. But this seems to us to be a response that itself evades what might be most challenging about these sorts of 'opinions' and 'attitudes', which we have seen to be often well-informed and always carefully reasoned. Perhaps what needs to be heard in this type of talk is the intuition being expressed that the ascription of *responsibility* to *consumers* is neither practically coherent nor normatively justifiable in quite the way that is assumed by many experts in policy, social science or activism.

5.5   Conclusion

This chapter has analysed talk generated from focus group research conducted in a diverse range of localities around Bristol in the UK. While we have not quite allowed our participants to just speak for themselves, we have tried to let their voices be heard. We have not interpreted their words against an external benchmark of what they should believe and how they should behave. We have tried to listen to what it is they are *doing* when they talk about not having the energy to bother about every choice they make when shopping, or not appreciating being made to feel guilty about so many of the world's problems, or just not agreeing that it is their responsibility to act more 'ethically'. Our focus group research demonstrates, first, that people from a variety of backgrounds have high levels of awareness of various issues related to the 'ethical' aspects of everyday consumption. Second, it shows that people engage critically and sceptically with the demands placed upon them as 'consumers' by campaigns and policies promoting ethical and responsible consumption. They do so by bringing a range of ethical concerns and competencies to their everyday consumption practices, ranging from the relatively personal responsibilities of family life to more public commitments like membership of particular faith communities, political parties and professional communities.

We want to underscore two themes that recur in these focus group discussions, both of which throw light on the ways in which the 'synapses' of consumer choice and ethical responsibility we discussed in Chapters 2, 3 and 4 are worked over by the practical reasoning of ordinary citizens. First, a great deal of everyday commodity consumption has little if anything to do with 'choice' as this is supposed to function by proponents of the market, left-liberal critics and grand sociological theory. Our research is consistent with Miller's (1998) argument that consumption is often embedded in networks of obligation, duty, sacrifice and love as well as in the ordinary, gendered work of social reproduction. Rather than aiming to be perfect ethical subjects, our participants displayed instead an aspiration to be 'good enough' moral subjects (see Smart and Neale 1997), assessing the demands placed upon them within the situated contexts of their everyday relationships and the valuations of worth present in these contexts. Even those participants who did consider themselves sympathetic to ethical consumption campaigns tended to have a modest and situated understanding of the consequences of their own actions in this sphere:

> CLAIRE: You like to think you are having an influence but at the end of the day I don't think so.
> ELLEN: I think of things like salad or mass-produced flowers for supermarkets and how little we are paying them to work, and pesticides, and illnesses. In Tesco the other day, they had fair trade roses and I thought actually I would be prepared to pay an extra pound for them.
> ALEX: More than thinking that I can change the world if I buy a certain way, I think I can influence the people around me. Maybe my friends will see that I have bought fair trade tea bags and the next time they are in the supermarket they think 'oh yes, that looks nice'.[28]

Our participants talked about consumption in a register of obligation and routine as much as they talked about 'choice'. When they did talk about 'choice', they very often expressed a sense that having to exercise so much energy over what to buy – on health grounds, on environmental grounds, out of a concern for oneself, one's loved ones, or distant strangers – was a burden. This in turn relates to the second theme to recur in these discussions. Our analysis illustrates that sometimes when people talk about their roles as consumers they accept that they do have certain responsibilities. Sometimes they do make *excuses* for not doing more. But very often they provide pertinent, informed, reasoned *justifications* for not considering it their responsibility at all. One finds people asserting finite limits to how much they, as individuals, can be practically expected to be responsible for; and one finds people articulating considerable scepticism towards the whole frame of 'individual responsibility' through which everyday consumption is publicly problematized.

In Chapter 2, we argued that ethical consumption campaigning should be conceptualized as a medium of ethical problematization, in which various discourses and technologies are publicly circulated in the attempt to position individuals as responsible subjects of their own actions. In Chapter 3 and Chapter 4, we argued that campaign organizations disseminate narrative storylines which seek to generate dilemmas around ordinary consumption activities. In short, these campaigns suppose that ordinary citizens are capable, practical reasoners. In this chapter, we have further developed this conceptualization of ethical problematization by exploring the ways in which ordinary citizens reason, practically, about the ethics of consumption. We have focused on elaborating the grammar of responsibility-talk around a topic, ethical consumption, which immediately generates a 'troubling' positioning for research participants. In section 5.3, we analysed discussions in which ethical products are described as too expensive, or as requiring too much time and energy, or being no fun, as 'factual versions' through which we can infer the meanings attributed to public discourses of ethical responsibility. These factual versions are the discursive mechanisms through which participants reason about the practical limits of demands made upon them. And in section 5.4, we analysed the *justifications* given by participants, the reasons they deploy to explain that 'ethical' demands are not necessarily more compelling than the obligations of everyday care relationships and ordinary routines. We have resisted the temptation to analyse these forms of talk as *excuses* precisely because rarely, if at all, are these utterances articulated in a register which implies that not acting 'ethically' is blameworthy. Quite the contrary, discussions of the difficulty of doing the right thing routinely shift to arguments about 'choice' being a burden or assertions that 'responsibility' really lies elsewhere.

In this chapter, adopting a theoretical and methodological approach which understands discourse to be a medium of accountability, we have argued that ordinary people engage in practical reasoning when confronted with the varied demands to consume 'ethically' which circulate publicly. As we argued in Chapter 3, this type of practical reasoning is a form of *action* that is an ordinary dimension of everyday life – it is part of the habitual coordination, maintenance and repair of relationships with people and things. There are three types of actions being performed in this type of responsibility-talk. First, when confronted with 'troubling' positionings that make moral demands on them, we see that people deploy factual versions to reflect on the practical limits of their capacity to act on these demands, versions which also reveal consumption to be embedded in personal and social practices that are loaded with ethical content of their own and difficult to adjust. Second, when confronted with 'troubling' positionings which make moral demands on them, we see that people also do relational reasoning, discussing how what they consider a binding demand on them is shaped by their obligations to others, close family members, or more anonymous social rules and institutions. And, third, when confronted with 'troubling' positionings which make moral demands on them, we see that people also engage in 'proper' deliberative reasoning – they work over examples, consider different values, give reasons and decide on the locus of responsibility for acting in response to collective matters of concern. In short, the ethical problematization of consumption around a register of global responsibility neither *subjects* individuals to a type of *disciplinary* force, nor does it merely *affect* their dispositions. Both of these dimensions of practice are folded up with deliberative practical reasoning of a very ordinary sort – this is presumed by the rationalities of ethical consumption campaigns, as we saw in Chapter 3 and Chapter 4, and this is evidenced by the analysis of how people talk about ethical consumption developed in this chapter.

The argument we developed in the first three chapters of this book is that the contemporary politicization of consumption involves efforts to encourage and enable ordinary people to articulate the background consumption built into their practices in relation to various normative schemas. We have argued that the concept of ethical problematization helps in understanding this trend: it points up the processes whereby practices are drawn into the orbit of discursive practices, are talked about, are made into fields of responsibility or personal choice. Problematizing consumption works through trying to attach various 'affective' dispositions such as love, compassion or sympathy to a new set of public concerns. But in this chapter, we have seen that people are quite capable of contesting the attribution to them of new global responsibilities. Attempts to reconfigure their affective attachments runs into various sorts of 'resistance'. The doubts and scepticism, the irony and the humour expressed by our focus group participants when we put them on the spot about the ethics of their own consumption behaviour need to be taken more seriously than they are by prevalent academic paradigms of both policy research and critical theory alike. What we hear in these discussions are people who are not well thought of as the ever-more individualized consumers of so much public policy discourse, nor as the dupes of neoliberal-consumerist hegemonies supposed by critical social science. They might be better thought of as little Habermasians, capable, willing and eager to engage in passionate citizenly discussion about where personal and private concerns rub up against political and public matters.

Chapter Six

Local Networks of Global Feeling

6.1 Locating the Fair Trade Consumer

In the previous chapter, we discussed the forms of ordinary reasoning through which the claims of ethical consumption campaigning are integrated into people's practices, or not. This chapter and the next one shift attention to one strand of ethical consumption campaigning, focusing on the fair trade movement. In looking at the dynamics of *consumption* of fairly traded goods, we seek to locate the analysis of ethical consumption practices in the context of debates about civic activism and political participation. There is an extensive literature which focuses on the relationship between the dynamics of social life and civic engagement and levels of activism. As we saw in Chapter 2, much of this literature presents consumerism as a corrosive social force that undermines the socio-cultural resources upon which active civic and public life depends. But we also saw in Chapter 2 that this type of analysis underestimates just how much civic and public engagement is actually generated by the social relations and material conditions of modern infrastructures of consumption. We have argued that if consumption is increasingly problematized as an object of public concern and political action, then the repertoires of consumerism have been creatively redeployed by activist and campaign organizations to mobilize consumption as a surface for the mobilization of attention and support for 'global' causes such as international trade justice, human rights and environmental sustainability.

In this chapter, we look at one example of how the problematization of consumption depends on existing networks of social relationships and civic association. We examine one aspect of fair trade consumption in the UK, focusing on how the mediating action of organizations and campaigns draws

Globalizing Responsibility: The Political Rationalities of Ethical Consumption by Clive Barnett, Paul Cloke, Nick Clarke and Alice Malpass © 2011 Clive Barnett, Paul Cloke, Nick Clarke and Alice Malpass

local social networks into campaigns which make claims on states, corporations and institutions at national and international levels. The argument is made by way of a case study of Traidcraft, a leading actor in the fair trade movement in the United Kingdom. The study focuses on how Traidcraft approaches and enrols its supporters, and how these supporters understand their own consumption and other practices. In examining what we call Traidcraft's political rationality, we focus on one of the two dimensions of social networks that Crossley (2007) identifies as important resources for the contemporary emergence of forms of extra-parliamentary politics. In this chapter, we examine the role of social networks and interpersonal ties in the *recruitment of individuals* into campaigns. In the next chapter, we examine the *block recruitment* of whole social networks into campaigns.

We use the case study of the organizational rationalities and practical actions articulated by Traidcraft to challenge the assumption that the actors involved in the growth of fair trade markets are best conceptualized as 'consumers'. The explicit aim of fair trade initiatives is to enhance democratization, empowerment and participation. This is widely acknowledged in literature on fair trade in the global South and in analyses of the impacts on producer communities (Bacon 2005; Doherty and Tranchell 2005; Leclair 2002). However, the same focus on the civic, political and social objectives of fair trade in the global North has been constrained by the view that the key actors in these practices are fair trade *consumers*. For example, Nicholls and Opal (2005) describe fair trade as a consumer choice movement driven by the forces of a market-based economy, and Wilkinson (2007) brands fair trade as a 'market-oriented social movement'. These characterizations of fair trade as using consumers and markets to pursue 'ethical' objectives sit uneasily with more radical interpretations that present fair trade as a political challenge to neoliberal commerce (Linton *et al.* 2004) and to the negative impacts of globalization (Raynolds *et al.* 2007). One of our aims in the next two chapters is to re-describe the dynamics of the growth of fair trade campaigning in such a way as to find a way around this persistent framing of the success of the fair trade movement as inevitably 'complicit' with larger macro-scale forces. We do so by seeking to describe the ways in which this movement reconfigures practices of political participation in the global North as much as in the global South, and to do so without presuming in advance that recourse to market-based, consumer-coded repertoires is incompatible with normative goals of global social justice.

In the next section, we argue that fair trade should be understood as a political phenomenon which, through the mediating action of organizations, coalitions and campaigns, makes claims on states, corporations and international institutions. Understood in this way, the growth of fair trade consumption in the global North works to mobilize support, raise funds and raise awareness about issues of global justice, development and inequality. In the third section, we look at how the fair trade movement seeks to mobilize support in the UK through various means, some directly related to consumption practices, some less so. Through the case study of Traidcraft, we shed light on the political rationality of fair trade consumption as it has developed in the UK. We show that it is far from obvious that the primary vector of engagement is individual's identities as consumers in the marketplace. The fourth section looks at how Traidcraft operates at grassroots level by drawing on existing social networks and civic associations to draw people into new arenas of public participation with global reach. We focus here on the understandings of participants themselves – of 'Fairtraders' and their customers – to demonstrate that the forms of subjectivity implicated in fair trade initiatives are more sociable than those suggested by automatic recourse to the ubiquitous figure of 'the consumer'.

6.2   Re-evaluating Fair Trade Consumption

This book draws on research undertaken in Bristol from 2003 to 2006 on various aspects of ethical consumption. A key component of this research focused on investigating various forms of fair trade campaigning in the city, and how these overlapped with other forms of ethical consumption and broader forms of civil participation and political action. Academic literatures in business, management and marketing studies, as well as fair trade organizations themselves, tend to distinguish 'fair trade' from 'ethical trade' (Smith and Barrientos 2005). Ethical trade focuses on labour conditions in mainstream production and distribution networks. Ethical trade campaigns display significant differences between, for example, the US and the UK (Hughes *et al.* 2007; Linton 2008). In the UK, the most visible of these campaigns is the Ethical Trading Initiative – an alliance of trade unions and NGOs focused on enforcing corporate codes of practice concerning working conditions and living wages in supply chains – whose corporate partners include major retailers such as Gap and Body Shop. In the United States, ethical trade campaigns have tended to be more fragmented organizationally.

Fair trade focuses on the development of alternative spaces of production, trade and consumption (Freidberg 2003; Raynolds 2000; Rice 2001). The international fair trade movement consists of certification agencies, producer organizations and cooperatives, trading networks and retailers. Since 2001, this movement has defined fair trade in the following way:

Fair trade is a trading partnership, based on dialogue, transparency, and respect, that seeks greater equity in international trade. It contributes to sustainable development by offering better trading conditions to, and securing the rights of, marginalized producers and workers – especially in the South. Fair trade organizations (backed by consumers) are engaged actively in supporting producers, awareness raising and campaigning for changes in the rules and practice of conventional international trade.[1]

The goals of fair trade include improving the livelihoods and well-being of producers 'by improving market access, strengthening producer organizations, paying a better price, and providing continuity in the trading relationship'; promoting development opportunities 'for disadvantaged producers, especially women and indigenous people'; awareness raising among consumers 'of the negative effects on producers of international trade so that they exercise their purchasing power positively'; providing an example of 'partnership in trade through dialogue, transparency, and respect'; campaigning 'for changes in the rules and practice of conventional international trade'; and protecting human rights 'by promoting social justice, sound environmental practices, and economic security' (Redfern and Snedker 2002). Fair trade organizations pursue these goals through trading activities, but also through awareness raising and campaigning.

Fair trade is a movement which seeks to harness the mechanisms of the market to address socio-economic inequalities and environmental harms associated with the global economic system (see Leclair 2002; Taylor *et al.* 2005; Young and Utting 2005). Fair trade involves three related activities: the organization of alternative trading networks; the marketing of Fairtrade-labelled goods through traders and retailers; and the campaign-based promotion of fair trade to lobby for changes to purchasing systems and global rules of trade (Raynolds *et al.* 2007; Wilkinson 2007; see also Lamb 2008). Leading commentators (e.g., Nicholls and Opal 2005) recognize four stages of history of the fair trade movement. First, following the Second World War, churches and charities in North America and Europe helped to bring relief to refugees by selling craftware in Northern markets. At the same time, Mennonite churches in the USA brought fairly traded embroidered goods from Puerto Rico into US markets. Secondly, alternative trading organizations such as GEPA (founded in Germany in 1975) and Traidcraft (founded in the UK in 1979) were formed to afford Southern producers the opportunity for direct trade with the North on an ethical basis. Such organizations were typically organized through church and community networks, and also retailed through 'One World' outlets and catalogues. Thirdly, fair trade retailing entered a period of considerable consolidation during the 1990s: alternative trading organizations launched mainstream product brands (such as *Cafédirect* coffee, *Divine* chocolate and *Geobar* cereal bars); key retail multiples (notably the Co-operative Group in the UK and Wild Oats Markets in the US) turned to fair trade as a significant part of their identity, thereby expanding the consumer base for fair trade products; and regulatory certification agencies such as the Fairtrade Foundation (founded in the UK in 1992), Transfair (founded in the USA in 1998) and the Fairtrade Labelling Organizations International (founded in 1997) were formed to provide benchmarked logos for fair trade brands. Finally, fair trade has most recently begun to find its place in the mainstream of retailing and consumption as fairly traded foodstuffs are increasingly marketed by leading supermarkets (for example, by Britain's largest retailer, and the third largest in the world, Tesco, which has launched own-brand fairly traded products), and as fair trade produce has become included in the product ranges of high-profile outlets (such as Starbucks and Costa Coffee). In addition, fair trade apparel is being now sold in mainstream clothes stores (e.g., Marks & Spencer in the UK) and other fair trade markets are being opened out, such as for sporting goods and jewellery.

Growth in fair trade markets has continued even in a period of economic downturn. According to the Fairtrade Foundation, an estimated €2.9 billion was spent on fair trade products in 2008, an increase of 22% over 2007. In the same period in the UK, fair trade sales exceeded £700 million, representing an annual growth of 43%. These increases during a period of global recession indicate a maturing market sector, although much of the increases referred to hinge on continuing growth in iconic fair trade product categories such as coffee, tea, bananas and cotton. By 2008 the global network of fair trade was reported by the Fairtrade Foundation to embrace 746 Fairtrade-certified producer organizations, representing over one and a half million farmers and workers worldwide.[2]

In the UK, the growth of fair trade consumption builds on a history of consumer-oriented campaigning by development and human rights organizations such as Oxfam and Amnesty. Oxfam's first charity shop was opened in Oxford in 1948, and there are now more than 750 Oxfam shops throughout the UK, selling second-hand books, clothes, records and other items, as well as (until very recently) a selection of fairly traded products such as tea, coffee and chocolate. Oxfam in turn is one part of a network of companies and organizations, including Cafédirect, Tropical Wholefoods and the Fairtrade Foundation, who have, over the past two decades, actively constructed networks for the production, distribution, marketing and retailing of fairly traded commodities. Traidcraft, the focus of our analysis here, is also part of this network, and reflects the wider influence of faith-motivated action in the fair trading arena (see Cloke et al. 2010). The network combines a range of organizations: some specializing in campaigning work, such as Oxfam; some specializing in marketing initiatives, such as the Fairtrade Foundation; others involved in business activities, such as Traidcraft – although these activities are not mutually exclusive. These organizations also have different forms of membership base: Oxfam's network of shops depends on volunteers; Traidcraft has an extensive network of volunteers buying and selling fair trade products in schools, churches and other non-commercial sites. This organizational framework is, in turn, embedded in broader institutional networks through which fair trade campaigning has been disseminated and from which support is drawn. This includes a history of fair trade activism in British religious organizations. More recently, fair trade campaign organizations such as the Fairtrade Foundation have cooperated with the leading labour union representative body in the UK, the Trade Union Congress, in seeking to ensure that *fair trade* activities support the *ethical trading* initiatives around labour rights of national and international labour union organizations.

In the European context, it is common to identify fair trade as part of a broader growth of so-called ethical consumerism or political consumerism (Bird and Hughes 1997; De Pelmacher *et al.* 2005; Hughes 2001; Nicholls 2002; Strong 1997). In political economy and critical social science, the predominant analytical frame for understanding both fair trade and ethical trade is that of commodity chains and value chains. This frame focuses on the links and connections between *producers* in the South and *consumers* in the North (Murray and Raynolds 2000; Raynolds 2002; Tallantire 2000; Tiffin 2002). There are three recurring topics of debate in academic literatures on fair trade. The first involves assessing the impacts of fair trading on producer communities (Bacon 2005; Moberg 2005), and seeks to acknowledge positive achievements without falling into overzealous depictions of fair trade outcomes (Hilson 2008). Fair trade does not function evenly across a range of different producer communities, with specific regional and historical contexts, and differing production units, partnership models and intricacies of community life ensuring that fair trade works differently in different settings (see Rice 2001; Tallontire 2000). As Ronchi (2002) argues, while fair trade umbrella bodies specify standards of behaviour for organizations, they do not stipulate any specific organizational model for fair trade, leading to a variety of organizational types, from small grassroots bodies, to consortia of small producers, to single cooperatives of large numbers of farmers. Since there is no model, there can be 'no fixed notions of what the impact [of fair trade] can be' (ibid.: 8).

Aside from this acknowledgement that fairness will mean different things in different producer communities, there is little reliable evidence on the overall impact of fair trade. In the absence of impact metrics of social accounting, which take account of producer community differences, we have to rely on the overviews provided by constituent organizations and on case studies of specific schemes. On this evidential basis, it appears that the direct influence of fair trade has transformed many producers' lives. Such transformation is partly about price differentials – Transfair USA estimate that in the first five years of fair trade activity in the USA, coffee farmers in the South received some US$30 million dollars more income than would have been the case at conventional prices (Transfair USA, 2004). However, transformation goes well beyond issues of trading prices. Case studies such as those by Blowfield and Gallet (2000), Kocken (2002), Malins and Blowfield (2000), Nelson and Galvez (2000) and Ronchi (2002, 2003) suggest a series of achievements of fair trade initiatives. For example, workers involved in producing fair trade products received higher incomes than those tied into conventional trade networks; particularly in cooperative settings, additional income is often pooled to be used in community education and health projects; some fair trade projects specifically seek to benefit and empower women; involvement in fair trade benefits producer identity and self-esteem; fair trade partnerships may have pioneered and promoted progressive work practices which have exemplified the benefits of high labour standards, worker participation and community support, and which set those standards across regions.

The second area of debate, building on the analytic frame of the chain and adding a culturalist dimension to it, revolves around processes of so-called *de-fetishization* of commodities. Work in this area argues that part of the success of the fair trade movement has been its ability to convince consumers that their purchases will be directly implicated in 'fairer' political ecologies of production, and therefore consumption (Bryant and Goodman 2004; Goodman 2004). Fair trade promotional materials and product packaging frequently present narratives which connect geographically situated farmers from the South with consumers from the North (e.g., Jackson 2004; Varul 2008) through the ethical work of a fair trade company. 'Fairness' is understood in these kinds of analysis to act as a kind of flexible and not quite empty signifier which helps to bind together and also demystify particular product chains and networks. Thus, it is argued that fair trade networks have the potential to reduce social distance (Moore 2004), and to stitch consumers into 'the very places and livelihood struggles of producers via embedded ethical, political and discursive networks' (Goodman 2004: 893), thereby serving to 'reconnect' consumers with producers in new forms of global moral economy. These types of cultural analysis of fair trade as a style of *consumerism* also generate ambivalent frames of evaluation: on the one hand, by bringing of moralized meaning into everyday lives the fair trade consumer is mobilized to respond to the hardships experienced by producers (see Goodman 2004; Hilson 2008); on the other hand, fair trade is presented as a means by which some consumers are presented with a mechanism with which to purchase a clear conscience via the encouragement of overconsumption (Renard 2003; Hudson and Hudson 2003).

The third area of debate around fair trade, one which also focuses on the ambivalent agency of 'consumers', is about mainstreaming. The initial development phases of fair trade were dominated by organizations formed specifically around the ethos of alternative trading methods and standards. Mainstreaming has involved the take up of fair trade as a trading niche by major multinationals whose *raison d'être* is fundamentally commercial and embedded in the free-market economy. Thus, in 2000, Transfair USA issued fair trade certification for Starbucks coffee, and in 2003 the FTF awarded fair trade status to products retailed by Tesco, which has subsequently developed a range of 'own-brand' fair trade products. Two principal fears have prevailed in this period of mainstreaming. The first starts off from the observation that, until recently, the growth of fair trade consumption has rested on a relationship of trust between consumers and retailers which has protected the value and meaning of 'fairness'. The concern is that the introduction of fair trade as a commercial niche by big business may erode the notion of fairness, as the small fair trade tributary becomes swamped in the flow of the corporate mainstream. As major supermarkets and corporations borrow or adopt the fair trade ethos, there is a concern that it will inevitably be set against conflicting commercial objectives (e.g., Ransom 2005).

The second fear about mainstreaming is the concern that corporations will undermine 'gold standard' certification conditions by using alternative fair trade regulators with lower standards. For example, the Kraft Foods Corporation (producers of Maxwell House and Kenco brands of coffee internationally) have recently announced a new line of products which carries with it certification from the Rainforest Alliance, a US-based NGO which seeks to ensure that coffee farms in Latin America are subject to rigorous environmental and social standards. The twinning of environmental with social concerns marks, for some, a welcome broadening of ethical concerns, but others see the proliferation of certifying agencies and goals as a route to diluting the standards of fairness involved. The risk is that commercially driven organizations will select a fair trade package that offers the highest gain in terms of perceived social responsibility at the lowest cost. Such practices would constitute a significant move away from the relationship of trust established by ethically motivated organizations and fair trade consumers.

Debates about mainstreaming also generate ambivalent frames of evaluation. Mainstreaming has been an essential element in the growth of fair trade, and in particular of market penetration by fair trade products (see Becchetti and Huybrechts 2008). At the same time, however, mainstreaming is seen to have 'bastardized' (Barrientos and Smith 2007: 57) the underlying concept of fair trade in that fair trade products are now being sold in contexts where other aspects of trading and business practice may be ethically questionable. This divided pattern of evaluation is a feature across all three of these areas of debate about fair trade production, consumerism and mainstreaming.

We want to interrupt this pattern of evaluation by calling into question the understandings of 'the consumer' and 'consumerism' that underwrite the analysis of fair trade consumption. We do so, first, in line with the argument developed in earlier chapters that ethical consumption often depends on the mobilization of identifications other than those of 'consumer', and, second, in order to better understand the contingent specificity of mobilizations of 'the consumer' when this is, indeed, the primary vector of identification engaged by campaigns. Positive evaluations of fair trade and other ethical consumer initiatives emphasize the potential of consumers to act creatively in the 'interstices' of the globalized economy in pursuit of environmental concerns, human rights or labour solidarity (Moore 2004; Renard 2003). There is a widespread assumption that the growth of fair trade is driven by and dependent on secular trends in consumer demand towards more 'ethical' and 'responsible' forms of product and service. This in turn generates the assumption that the growth, potential and limits of fair trade are determined by the dynamics of consumer demand, understood in terms of the individual choices of rational utility-maximizers. Green or ethical consumers are seen as the drivers of market change towards more virtuous or fair systems through the mechanism of informed choices. The citizen-consumer can, on this interpretation, be seen as acting as an agent of regulation through the market, stepping into the vacuum left by the apparent retreat of state actors from this function.

This positive view of the role of citizen-consumers is mirrored by a more critical attitude to consumer-oriented forms of social action. The more pessimistic view of the potential of 'shopping to change the planet' (Seyfang 2004) sees the growth of ethical consumerism, sustainable consumption initiatives and fair trade as parasitic upon and further contributing to a thoroughgoing individualization of civic, public culture, which legitimizes the hollowing-out of the responsibilities and accountabilities of nation-states. Low and Davenport (2005) argue that the mainstreaming of fair trade has given impetus to the growth of discourses of ethical consumption that undermine projects of collective mobilization. On this view, ethical consumerism is an individualizing phenomenon which lacks a collective dimension (Seyfang 2005). It empowers some – those with the spending power to make their 'vote' in the marketplace effective – at the expense of the egalitarianism of formal, public, representative institutions. The 'alternative' potentials of fair trade are, on this interpretation, limited by the reliance on the conventions and codes of (post)modern consumerism which, in the final analysis, are complicit with the cultural logics of market-driven capital accumulation and the crisis of democratic accountability (Bryant and Goodman 2004; Goodman 2004). Such a critique of the articulation of consumerist discourse with campaigns against global poverty and in support of trade justice and environmental sustainability remains wedded to an image of the consumer as the key agent of market change. In this representation, the people who buy and sell fair trade products in the North remain curiously abstract, detached, even placeless actors.

The focus on consumer demand as the driving force of the growth of fair trade suggests that the only objective pursued by the fair trade movement is further economic expansion. For example, the development potential of international fair trade is seen to depend on finding ways of maintaining and expanding consumer demand (Levi and Linton 2003); and innovative marketing strategies are thought to be the best mechanism for taking advantage of market opportunities (e.g., Golding and Peattie 2005; Randall 2005). This in turn generates an evaluative framework in which the problem of 'mainstreaming' is conceptualized in narrowly economic terms: does the shift from 'niche' to 'mainstream' retailing threaten to the reabsorb original aims into market logic? (see e.g., Hira and Ferrie 2006; Renard 2005; P. Taylor 2005).

Low and Davenport (2005) go so far as to suggest that one effect of the mainstreaming of fair trade is that, as they put it, the medium gets confused with the message. By this, they mean that 'shopping for change' increasingly comes to be seen as the primary mechanism of socio-economic change, rather than as one aspect of a broader movement focused on trade reform and trade justice. And, in turn, the 'dilution' of the ideals of fair trade is always already inscribed within the growth of fair trade or similar alternative economic practices as long as they are conceptualized as primarily economic means of achieving economic objectives.

Our argument is that the assumption that the key driver of the expansion of fair trade consumption is 'the consumer' needs to be called into question. As we have already argued in previous chapters, when looked at in more detail it turns out that ethical consumption campaigning in the global North does not seek to mobilize people as individualistic consumers. In contrast to the view of ethical consumption as a substitute for other forms of more collective engagement, we argue here that there is a spectrum of actions through which consumption is problematized as both an object and medium of ethical commitment and political participation. This spectrum ranges from more individualized, discrete activities such as purchasing fairly traded products in the supermarket, to more sociable practices such as involvement in local campaigns to have schools, universities or towns certified as 'fair trade', through to explicitly political engagement, whether through individualized petition-signing or collective involvement in mass demonstrations. Even the most individualized and consumerist of these activities is, then, connected to a broader range of actions. This connection is sometimes directly undertaken by the same person who buys fair trade products – they might also be signing petitions, donating money, boycotting other goods, joining local campaigns, attending meetings or going on marches. We will explore this repertoire of actions further below. But even in the case of anonymous consumers who are responsible for the steady growth in fair trade consumption in the UK, it is important to recognize that this market for fair trading is not a spontaneous response to 'revealed preferences'. In this chapter, we argue that the possibility of exercising individual choice as a *fair trade consumer*, or any other type of ethical consumer for that matter, is made possible by intermediary trading, marketing, campaigning and educational organizations with diverse modes of membership and support in civil society. Furthermore, the growth of fair trade retail markets in the UK is closely associated with a pluralization of campaigning strategies by fair trade organizations. Far from focusing solely on growing commercial retail markets in fair trade, it is increasingly the case that fair trade initiatives are aimed at transforming infrastructures of collective consumption through which the 'choice sets' available to consumers are realigned to support fair trade objectives. In short, the growth of markets for fair trade products has been associated with the development of new forms of collective, organized political action that remains focused on questions of poverty, sustainability and justice, and which takes states, international agencies, and multinational corporations as objects of contention. It is therefore important to acknowledge 'the ways in which the fair trade movement encourages actors to engage in different forms of social action' (Shreck 2005: 18). And in order to fully develop the insight that fair trade facilitates participatory forms of social action, the analytical focus on 'consumers' as agents of fair trade consumption needs to be displaced in order to better understand the contingent mobilization of these figures in fair trade consumption campaigning.

6.3   Managing Fair Trade, Mobilizing Networks

We have argued in previous chapters that interventions in the politics of consumption start from the recognition of the complex folding of commodity consumption into affective and material infrastructures of everyday life. This reflects the recognition that levels of commodity consumption are not straightforwardly sustained by consumer demand, and that therefore 'consumers' are not straightforwardly the most effective agents of change. We saw in Chapters 3 and 5 that a great deal of consumption is embedded in practices where people are acting as parents, caring partners, football fans, or good friends. Some consumption is used to sustain these sorts of relationships: giving gifts, buying school lunches, getting hold of this season's new strip. And Chapters 3 and 4 discussed the ways in which a lot of consumption is done as the background to these sorts of sociable activities, embedded in all sorts of infrastructures (transport, energy, water) over which people have little or no direct influence as 'consumers'. In this section, we want to further develop the sense that people engage in ethical consumption through identifications which extend beyond the thinness of 'the consumer' by examining how social networks and interpersonal ties are used in the recruitment of individuals into fair trade campaigns.

Fair trade campaigning in the UK is one example of this emergent problematization of consumption as a function of practices and routines. The fair trade movement began to develop internationally in the 1980s, with the launch of 'Fairtrade' labelling in various national contexts, and in the 1990s the emergence of international umbrella organizations establishing worldwide standards of labelling and certification. In the UK, the Fairtrade Foundation (FTF) is the NGO that certifies the use of its distinctive *FAIRTRADE* mark in accordance with this international network. Set up in 1992 by leading development charities (CAFOD, Christian Aid, Oxfam, Traidcraft, and the World Development Movement),

FTF uses its certification process as a means of enlisting support, raising awareness and transforming consumption practices in line with the principles of fair trade. The campaign devices it deploys have included the extension of Fairtrade certification beyond commodities to various institutional actors such as schools, churches, universities and even localities. The evolution of Fairtrade certification illustrates that, sometimes, campaigning on fair trade deploys devices that are presented as extending choices to consumers. These are primarily used as a means of raising awareness among a broad general public and generating media attention. But campaigns also engage at an institutional level to change the ways in which consumption is regulated at the level of whole systems of provisioning. This combination indicates that fair trade networks succeed not by linking 'Third World producers' with placeless 'First World consumers', but by articulating social networks and their members into new transnational geographies of place.

The development of Fairtrade certification in the UK illustrates the need to resituate the analysis of fair trade consumption in diverse social networks, rather than persisting in presenting it (positively or negatively) as primarily a form of consumer agency. To this end, we present here a case study of fair trade networks in and around the city of Bristol in the south-west of England. We focus on two distinct but overlapping local networks of activism, one centred on church-based networks in North Somerset and South Bristol, the other linked closely to networks in South Gloucestershire and North Bristol. We draw on empirical research into the role of Traidcraft, one of the UK's leading fair trade advocacy organizations, undertaken between Autumn 2003 and Spring 2006. Primary data collection included desk-based investigation into Traidcraft's development, using the organization's archive at its head office in Gateshead in north-east England, and interviews with key actors in Traidcraft's national management, policy, and business operations. It also involved in-depth, semi-structured interviews with local 'Fairtraders' and Traidcraft customers. This research into local Traidcraft networks used the organization's three local 'Key Contacts' in Bristol to recruit 15 further research subjects for in-depth interviews.

Traidcraft was formed in 1979, emerging from the Christian development charity Tearfund. It is a trading business whose founding principles are to provide a Christian response to poverty through trade. It was set up to distribute fairly traded food, household and craft products throughout the UK. Traidcraft's initial system for distributing products was through a series of local representatives – 'Fairtraders' – who volunteered to sell stocks of fair trade goods in their area, as well as to distribute mail order catalogues and generally raise awareness of the issues surrounding fair trade. There are now some 5,000 volunteers in Traidcraft's national network. 'Fairtraders' are not, it should be noted, straightforwardly 'consumers': they buy fair trade products from Traidcraft, and then sell them on through their own social networks, enrolling friends, parishioners and work colleagues as consumers in more or less inadvertent ways.

While Christian principles remain central to Traidcraft's operations, it is formally committed to working with people and organizations from any faith background, or none at all (Moore and Beadle 2006). As fair trade markets have grown, Traidcraft has become a leading fair trade brand in the UK. It has mainstreamed its distribution through involvement with fair trade product lines such as Cafédirect and GeoBars that are marketed through high street retailers and supermarkets. As locally based 'fair trading' has grown, Traidcraft has established a number of Key Contacts across many localities, who provide wholesaling and information services for Fairtraders working at ground level in churches, schools and workplaces. In 2005, Traidcraft sourced craft, food and clothing products from approximately 100 producer groups in more than 30 countries, and sold them in the UK through almost 5,000 volunteers. These volunteers bulk buy products to sell to friends, neighbours, work colleagues and fellow churchgoers (accounting for about half of all sales), and also through supermarkets and wholesalers, independent retailers and mail-order/e-commerce distributors (accounting for the other half).

The development of Traidcraft's trading business practices over time has been associated with organizational shifts in its charitable and political activities. It is a founder member of the International Fairtrade Association and the European Fairtrade Association, and in 1992 was one of the founding organizations behind the formation of the Fairtrade Foundation (FTF) in the UK. Traidcraft is not, then, simply a business organization. It aims to operate at three levels: building up trade directly; developing an infrastructure that can deliver business development services and support to marginalized groups; and working at a policy level to help governments support poverty-reducing trade and to help governments in international trade negotiations.[3] Given the multidimensional nature of Traidcraft's activities, it is important to acknowledge its role in networks of transnational advocacy around issues of development, global poverty and trade justice in addition to its status as a model of alternative trading or a paradigm of the virtuous business organization. This shift of perspective reveals that Traidcraft's guiding rationality does not simply reproduce the notion that the power of markets can serve as substitutes for concerted collective action by states, international agencies, NGOs, or social movement organizations if brought under the sway of appropriately virtuous 'consumer choice'. Rather, Traidcraft functions as one organization within a broader set of networks which provide pathways through which relatively individualized actions are articulated with more collectivized modes of action and, in the process, help to shape new forms of public life, develop new market practices and promote alternative visions of economic futures.

Traidcraft's role as an intermediary actor in networks of civic activism and political participation can be helpfully analysed along the lines suggested by Pattie *et al.* (2003) in their extensive analysis of civic life in the UK. They distinguish between three types of 'activism': individualistic activism (e.g., wearing a campaign badge, buying fairly traded coffee); contact activism (e.g., writing to a Member of Parliament, signing a petition); and collective activism (e.g., attending a public demonstration, joining a trade union). Traidcraft presents its activities along each of these three dimensions, coinciding with its distinct organizational division between its trading activities, its charitable activities and its lobbying and campaigning activities. Each of these three organizational dimensions is translated into a simple mode of personal action that can be easily undertaken by ordinary people:

> You buy – and we can trade; you donate – and we can support; you campaign – and we can influence.[4]

The 'buying' aspect of this formula is managed through Traidcraft plc, the trading arm of The Traidcraft Foundation. The 'giving' aspect is related to the work of Traidcraft Exchange, founded in 1986 to generate charitable funding to support capacity building work in the global South and awareness raising in the global North. When people fill out a catalogue order with Traidcraft plc, for example, they can elect to make a donation to Traidcraft Exchange at the same time. These dimensions of Traidcraft's operations therefore facilitate forms of 'individualistic activism'. But Traidcraft also provides opportunities for Fairtraders or fair trade customers to engage in forms of 'contact' and 'collective' activism too. Via its web site and regular mailings, it encourages people to express support for campaigns through formatted emails, postcards and letters. The Traidcraft Speaker Scheme provides volunteers with the opportunity to present at school assemblies or in the classroom, to give sermons in churches, or to appear on local radio or news. And it provides a number of pedagogic resources to assist these volunteers, including interactive games, videos and DVDs, maps, information sheets, prayers and sermon outlines.

These forms of mundane 'contact' and 'collective' activism are closely related to the work of Traidcraft's Policy Unit, established in 1998. This formalization of Traidcraft's advocacy and lobbying activities followed in part from the development of the organization's internal financial resources. It also reflected a response to the new opportunities for access to government following the election of a Labour government in 1997; for example, Traidcraft quickly became part of the Department of Trade and Industry's Trade Policy Consultative Forum. But the policy and advocacy activities of Traidcraft are also indicative of a long-standing acknowledgement that 'one relatively small trading organization cannot solve the problems of the world on its own'.[5] In both the UK and at the EU level, Traidcraft has succeeded in gaining 'standing' as one representative organization with a voice on trade issues. It is also part of networks involved in lobbying policy makers and legislators. For example, Traidcraft is part of the steering group of the Corporate Responsibility (CORE) Coalition set up in 2001, and a member organization of the Trade Justice Movement (TJM) formed in 2000. In 2005 and 2006, CORE and TJM coordinated a media and lobbying campaign to amend the Company Reform Law Bill during its passage through parliament. This campaign focuses on demands that UK companies be made legally accountable for their activities overseas, through mandatory auditing and monitoring procedures, and by allowing non-UK citizens and residents to take legal action in UK courts to seek redress for harm and injustice caused by UK-based companies in their own countries. This campaign is just one example of the increasingly prevalent strategic phenomena of NGO coalition-formation around particular issues, drawing together organizations focused on development, human rights, trade justice and environment into contingent networks of advocacy (Yanacopulos 2005). This sort of advocacy at national and transnational scales is, though, sustained by more localized organized advocacy. Traidcraft is an organization that can draw on a committed support base to represent the organization in various local and regional networks, such as the South Gloucestershire Fairtrade Network or the Bristol Trade Justice Group in our case study. It is through these networks that volunteers come to participate in numerous forms of collective action, including organizing Jubilee 2000 debt campaign events or attending public demonstrations during the Make Poverty History campaign of 2005.

Traidcraft's articulation of its volunteer network with national and transnational coalitions illustrates a broader political rationality operating in policy and advocacy fields concerned with issues of consumption, markets, poverty and sustainability. As we saw in Chapters 3 and 4, think-tanks and advocacy organizations increasingly argue that the key to influencing consumption practices is to intervene in processes of shared learning through peer groups and social networks. This political rationality recognizes that people's motivations as 'consumers' are not necessarily individualized at all, but are embedded in networks of sociability. This is indicated by the recognition that transforming practices depends on the 'discursive elaboration' of existing habits and routines (Burgess 2003). This understanding is already well established in ethical consumer activism, and is exemplified in the case of fair trade by Traidcraft. Rather than seeking to change people's opinions and preferences, Traidcraft seeks to *extend* people's existing dispositions into new areas:

STUART PALMER: We used to try and lead people to think and act in particular ways. Now we try and respond to people and provide outlets for their energy and commitments.[6]

1. Do you and your family buy Fairtrade products?
2. Do you believe actions speak louder than words?
3. Are you keen on meeting like-minded individuals?
4. Are you interested in getting or staying fit?
5. Do you enjoy walking, running, cycling?
6. Have you travelled off the beaten path?
7. Would you like to get involved in fighting global poverty?

If you answered 'yes' to questions 1 and 7, and 'yes' to one or more of questions 2 through 6, then congratulations, you fit the profile of a Traidcraft GeoActivist!

Figure 6.1   Traidcraft GeoActivist Personality Test (*Source*: www.traidcraftinteractive.org.uk/news.php?story=46, accessed 21 August 2006)

An example of this focus on developing people's 'energies and commitments' is Traidcraft's *GeoActivist* initiative. The initiative is led by an interactive web site[7], which includes the GeoActivist Personality Test.

The GeoActivist Personality Test (see Figure 6.1) is an example of Internet communication being used as more than a source of information, but as a campaigning device (Kleine 2005). The GeoActivist initiative seeks to enrol 'healthy lifestyle advocates, walkers, runners, cyclists, world explorers, and fair trade activists' as fundraisers, through organized sponsored walks, rides and runs, thereby connecting existing leisure and lifestyle commitments to concerns with trade justice and global poverty. It is an example of Traidcraft's sensitivity as an organization to the differentiated practices, resources, commitments and concerns that lead people to support the principles of fair trade. It also reflects an explicit concern to widen its support base by hooking into existing networks:

BRIAN CONDER: We have the GeoActivists, which is quite new and is used for raising our profile at various sporting events such as the Great North Run, the Great South Run, etc. So that is one way of doing it. But also, just when we target churches we will specifically target the material so that it will not necessarily lean away from the older person, but the graphics that we show on there might be showing younger people at work or working with children at the church to help promote the product. We are getting very involved in school groups now with our Young Cooperative scheme, which is aimed at middle schools and upwards. Ministers of religion are one of our targets over the next few years to increase the number of Fairtraders from church-based backgrounds. So we will target ministers, who again we will explain to them what we are about and they will hopefully explain to their congregation what we are about. It is not specifically aimed at the younger people but a lot of the materials that we produce will actually give that impression if you like.[8]

This account of the variety of strategies for enrolling supporters illustrates the organizational focus on using key intermediaries (e.g., ministers of religion) and institutional sites (e.g., schools) to enlist individuals into participatory practices in order to sustain and grow Traidcraft's volunteer network. It underscores the argument we are developing here: the development of fair trade consumption in the UK does not aim simply to enrol people as 'consumers'. It addresses them as members of varied social networks with the aim of extending their commitments into their consumption habits and channelling their energies into recruiting friends, family, work colleagues, or fellow parishioners.

Traidcraft remains most active in church-based networks, and churches are important sites for recruitment through interpersonal ties. Churches are key 'spaces of fair trade' in the UK, perhaps only matched by Oxfam shops. There are widespread variations in levels of fair trade activism between churches and denominations, but many churches provide social spaces through which new customers, Fairtraders, and supporters can be recruited, and through which fair trade can be diffused into other social and institutional sites in local areas – schools, clubs, unions. Amongst the respondents in our case study, it was a common experience for people to have first come across the fair trade movement through interactions with Traidcraft representatives who belonged to the same church networks. Traidcraft recognizes that word-of-mouth is a basic way in which fair trade networks extend into existing social networks and civic associations: 'The majority of our new Fairtraders come through word-of-mouth recommendation'.[9] One local Fairtrader described this as a basic tactic she has used herself:

EMMA: I think it's certainly the best way of spreading their message by getting people talking to local people to then spread it further. It's kind of, it's filtering out again, it's branching out. If they only operated through catalogues I think it would be, they would have much less support, because a lot of it is word-of-mouth and friends doing this and so-and-so talking about that, and just seeing the products and being able to get them easily and make, and them being easily accessible. I think people then do see, would try something from Traidcraft on a much more casual basis than they would if they had to then order it from a catalogue or make any commitment or something and pay post and packing and all that sort of thing.[10]

Identifying and recruiting key intermediaries and facilitating social networking are central objectives of the organizational rationality of Traidcraft as it has developed and grown. In the next section, we look at how these mechanisms of enrolment operate locally. This will reveal that local fair trade activism is not only oriented to the economic objective of growing fair trade markets, but is instrumental in the development of multidimensional patterns of political action into which Traidcraft and its Fairtraders are woven.

6.4   Doing Fair Trade: Buying, Giving, Campaigning

The previous section showed that strategies to extend fair trade consumption do not necessarily aim to enrol people as 'consumers'. They are just as likely to recruit supporters through interpersonal ties, embedded in spaces of social interaction such as churches, schools or workplaces. In this section, we show that people who engage actively with fair trade consumption activities similarly do not do so by recognizing themselves as 'consumers', but rather use consumption practices to express existing commitments to various ethical and political projects. The focus on 'the consumer' as the key agent in fair trade consumption tends to hide from view the relations between fair trade consumption and other forms of civic and political action. This section looks at two overlapping networks of local fair trade activism in the south-west of England. We examine the way in which participants in these networks understand their own activities, and find that various identities are at work, but rarely if ever that of 'consumer'.

As an organization rooted in Christian principles, churches and faith-communities are the key social networks in which Traidcraft is embedded:

> BRIAN CONDER: The majority of Fairtraders sell in church and a number of their customers will be fellow churchgoers and that will range from people who are very, very much wanting to buy the product and will buy it regularly in fair-sized quantities, to other people who will see the stall at the back of the church and think I have to do my little bit and they will buy a jar of coffee and nothing else.[11]

It is to be expected, then, that amongst participants in Traidcraft networks, the motivation for buying or selling fair trade products was often closely wrapped up with individuals' Christian faith commitments. For many of the research subjects in our study (although not all), doing fair trade represents one route through which they feel able to express the integrity of their faith identities in everyday ways (Coşgel and Minkler 2004). However, our research subjects tended to reflect on the relationship between their faith and their involvement in fair trade in a particular discursive register. While often acknowledging that for them personally, faith and church membership were important factors in their fair trade consumption activities, they just as often insisted that this was not a necessary relationship, either for them or others. For example, Sarah, a Fairtrade customer from North Bristol, felt that: 'It just seems the moral thing to do. I just believe in the ethos of what it's about'.[12] Likewise, Liz, a fair trade customer in North Somerset, acknowledged the importance of her involvement in the church, but insisted: 'But that's not why I support fair trade. I do it for ethical reasons. I think it's a really good idea'.[13] She felt that ethical reasons and faith motivations were not the same thing:

LIZ: How is it different? Well if I didn't go to, if I wasn't a Christian, I would still want to do it, yes, because I do care about the world and everyone in it, I don't just think about my own family, my own village, my own country. I think we should help people in the Third World, and it's not right that a lot of people are making money at their expense [. . .]. If I wasn't a Christian I would still support fair trade, yes, because I have a social conscience and I care about people and I don't think its right that they should be taken advantage of.[14]

Another respondent, a fair trade customer and activist in local schools, expressed a similar view:

ERIKA: No, I will support any organization. My own motivation might be my faith, support might be Christian, but anybody with an interest in humanity I suppose; any faith I think would be; I would support just about as much. It's the outcome that counts. I'm not a missionary.[15]

These reflections on the relationship between personal faith, church member-ship and involvement in fair trade networks indicate that participants in local fair trade activities have a strong sense of the *contingency* of this relationship between faith and fair trade. These respondents recognize that the relationship between fair trade and church membership or personal faith is not a natural, necessary, or automatic one. There are two aspects of this recognition. It partly reflects a sensitivity about over-identifying fair trade with Christian faith:

ERIKA: Movements that are very ideologically based, or too faith – faith in that sense being an ideology I suppose – they can be quite inflexible and become a bit strident later, where once something reaches a certain momentum, it may be more practical and more useful to have a bit more realism and compromise.[16]

Second, the acknowledgment of the contingency of the relationship between faith and fair trade also reflects the experience of many of our research sub-jects of having themselves had to work hard amongst their own church communities to encourage the adoption of fair trade practices. The recourse to the discourse of faith is a strategic option that can serve a useful purpose in persuading other churchgoers to support fair trade. For Traidcraft repre-sentatives and committed Fairtraders, a shared discourse of faith can often provide an effective way of translating their personal commitments to fair trade and trade justice into these institutional social spaces, in which an overtly 'political' address might not necessarily succeed:

SUE: Well, I don't think you have to be a Christian to want to support fair trade, but that was the way I persuaded the people at the church that it was the right thing to do, because I think it was easier for them to come to terms with it, looking at it in that context, than as a wider issue. I mean, there are political arguments for doing it too, I feel, but that wasn't the way to persuade them, so I didn't bother with that.[17]

Sue's reflections indicate that it should not be assumed that the fair trade movement has a natural home in church-based networks. These participants have a strong sense of the hard work involved in the active transformation of these networks into distinctive, novel communities of 'global feeling' focused on issues of trade justice and global poverty. Like Sue, Edna has harnessed Christian discourse when appropriate to persuade people to support fair trade, downplaying the 'political' aspects because ' "political" was a bit of a dirty word within the church':

EDNA: We were looked on a little bit with some sort of suspicion. It was very difficult to get fair trade into a lot of churches.[18]

At the same time, Fairtraders are keen to extend fair trade beyond the social spaces of the church, and sometimes even keen to loosen the association between fair trade and Christian faith.

If faith commitment plays a specific role in motivating the involvements of members of these fair trade networks, then in turn their involvement in fair trade consumption is not strongly motivated by an abstract sense of 'consumer power'. They are quite realistic about the limits of consumer activism undertaken in isolation from other activities. Their involvement tends to follow as an adjunct of thicker forms of identification and modes of sociability. This finding is underscored by the acute sense of the potential and limits of fair trade to transform patterns of global trade on its own. Our respondents had few illusions about the magical power of consumer demand to shift markets, corporations or governments. Debbie, one of Traidcraft's Key Contacts in Bristol, sounded almost pessimistic on this score:

DEBBIE: I could be wrong. I'd like to be wrong. I don't think it will actually overturn the desires and aspirations of the corporations. I think they've got at the moment far too strong a grip, through governments.[19]

This view seems to beg the question of why people would engage in fair trade buying and selling in the first place if it is not to contribute to changing global markets. Sarah provides a nuanced sense of why she felt it worth while to do so:

SARAH: Consumers give their voice, don't they, by what they buy? They're showing, even the big supermarkets, they're showing them what you want them to stock, aren't you, by buying these things? So, yeah, I can see that by buying Traidcraft you're showing that.[20]

The emphasis, for Sarah, is not so much on the aggregate consequences of purchasing decisions, but rather the sense that fair trade consumption is an avenue for expressing one's commitments, of having them registered by powerful actors in the public realm.

Another Fairtrader provided a more sustained justification of fair trade consumption. This centred on fair trade consumption as a medium for establishing and maintaining a sense of agency amongst ordinary people:

> HATTIE: The thing that you have to break through is the 'It's such a big prob-lem, what can I do about it?'. Yeah, okay, if a government can do something about it then that's great, if the international financial facility happens and aid doubles then that would be wonderful, but people feel that they can't indi-vidually do anything – whereas, in fact, fair trade is something you can do individually.[21]

The difference that fair trade can make from Hattie's perspective, her reason for being involved, does not lie in her individual actions. It lies in being involved in a practice that demonstrates a sense of empowerment through concerted action:

> HATTIE: Because if everyone thinks they can't make a difference then they don't, but if each person just changes one or two things that they buy, and it just begins to build, and it is something that you can do and it is something one person can do, that a group of people in a village or a town can make a difference to.[22]

For Hattie, the success of fair trade in the market demonstrates this possi-bility of 'making a difference':

> HATTIE: And it does add up. I think it's that thing of 'Yes, individual action can actually make a difference and add in', and that's really important in lots of ways, because that's where people feel disempowered and disassociated from politics with a capital 'P' and think 'what I do doesn't make any differ-ence' or 'what I say nobody listens to'. And the more people see that it does, the better.[23]

This ordinary understanding amongst both Fairtraders and their cus-tomers is indicative of an appreciation of markets as kinds of public space, in which values can be expressed, the existence of certain constituencies made visible, and positive capacities affirmed. It also indicates a clear sense of realism about the position of fair trade as one part of a broader move-ment focused on trade justice, labour solidarity, and human rights, as expressed by another of our respondents:

> JOHN: Fair trade per se is tiny, it's grown a lot recently, it's still minuscule compared to world trade as a whole, but I think its very important to have

standards set, saying yes you can trade ethically. You're waving the flag, which has much wider influence on other organizations, and there's things like the Ethical Trading Initiative which affects the big boys, and I think that's very driven by the existence of full-on fair trade. I think that it's an ethical driver.[24]

John's sense of fair trade as an 'ethical driver', as enacting a kind of demonstration effect, throws light upon the problem of mainstreaming that recurs throughout discussions of fair trade and ethical consumerism. As we suggested in the first two sections, the narrowly economistic understanding of fair trade tends to offset the virtue of fair trade's origins in niche markets to the hazards of expanding into mainstream retailing. For fair trade activists, though, mainstreaming is not necessarily seen as a problem in this way at all. Pauline, a fair trade customer who is active in the Trade Justice Movement and a member of Christian Aid, actually goes out of her way to buy fair trade products in supermarkets:

> PAULINE: If it's available in a supermarket, I would rather buy from the supermarket. I would rather buy from the supermarket because they look through all their sales, don't they, keep all that data, and if they can see that people, that there's people buying the Fairtrade stuff or Traidcraft stuff, they're gonna stock more of it, so where possible I would buy from the supermarket.[25]

This view is, of course, quite consistent with the primary objective of generating benefits for producer communities through increasing the market for fair trade. But Pauline also expresses an understanding that the growth of fair trade in mainstream retailing is a further boon to raising public awareness of trade justice issues. And this view is consistent with Traidcraft's own involvement in advocacy and lobbying:

> BRIAN CONDOR: I mean, the contacts that we have now with movers and influencers have grown alongside a growth in fair trade more generally and our customers. The more mainstream it becomes, the more opportunities are created for people at our Policy Unit to talk to people and to change opinions[26]

Not only does mainstreaming provide opportunities for further awareness raising, but it also shows a level of support among 'consumers' whom Traidcraft can claim to represent. As we argued in Chapter 4, when engaging in policy arenas with government officials, regulators and business, this capacity to speak authoritatively for 'the consumer' carries important persuasive force. Sales figures for Traidcraft plc and donation figures for Traidcraft Exchange are important devices in enabling this representative work.[27] This strategic use of sales figures in lobbying and campaigning points up one of the most interesting facets of fair trade consumption as a 'surface of mobilization': while a great deal of fair trade consumption is not motivated by the sovereign acts of rationalizing consumers, these acts of purchasing are counted and reported by organizations as if they were reflections of consumer preferences. Being able to speak in the name of 'the consumer' is, for these organizations, a crucial means of representing the popular will in engagements with business, regulators and governments.

For our informants in and around Bristol, buying or selling fair trade was a natural extension into the everyday routines of social reproduction of the commitments they already sustained in other civic, community and political practices. Our research subjects expressed a range of understandings of whether and how their fair trade activities counted as 'political' or not. Sarah, whose involvement in fair trade is only as a customer, sees this as 'Just doing my little bit I suppose. It's that "think global act local" I suppose, that sort of thing'.[28] Likewise, Liz, another Traidcraft customer, does not involve herself in public campaigns with which Traidcraft is associated, such as Make Poverty History, 'because I felt I supported it in other ways, through supporting fair trade'.[29] But she acknowledges the facilitating role of Fairtraders and Key Contacts in enabling her to be part of these networks:

> LIZ: They make it possible, yes. Because I'm not the sort of person, I admire all these young people who usually are, who go out and do things and go out and help in a crisis. I don't think I would ever have been, I'm not the sort of person, I wouldn't feel brave enough to do it myself, I would rather help in other ways.[30]

Liz sees her activity not as 'political' but as 'practical'.

Liz is one of the customers in North Somerset supplied by Debbie, one of Bristol's Key Contacts. Debbie does recognize her activity as a kind of political action:

> DEBBIE: Yes I do, but I mean in the actual taking action, going out on the streets and going to rallies, that type of thing, and meetings as to what are we going to do, how are we going to approach this campaign, I don't go to the campaign meetings and I don't go out on the demos [. . .]. I mean I'd love to, I'd love to, but I don't have time. So on the practical side I feel I'm there to provide the stock, provide myself if that's required, and provide information if they want it, or I can point them in the direction to go.[31]

Debbie describes herself as 'a networker' and a 'frustrated activist'. Her own reason for not engaging in more formal campaigning is because she invests so much time enrolling people into consumption of fair trade products. Debbie literally 'sets out her stall' in all sorts of mundane spaces: churches, village fetes, youth clubs, schools, Women's Institute meetings. This type of practical advocacy ensures that fair trade goods and literature can be made present in these sorts of spaces and on these sorts of occasions: she is an agent for the low-level diffusion of fair trade products and discourses into local social networks.

Beyond those for whom their main involvement in local fair trade networks is as a customer-supporter, and those like Debbie who actively work to sustain and extend these networks of fair trade supporters, one finds people in these local networks for whom fair trade consumption is just one part of a much more explicitly political self-identity. John combines purchasing fair trade products with other ethical consumption practices: he boycotts Esso, buys organic food, and invests with an ethical bank, Triodos. He is also an active fair trade campaigner at work, persuading colleagues to switch to fair trade coffee; a member of the Labour Party and Friends of the Earth; and has been actively involved in advocacy work with the Ecumenical Council on Corporate Social Responsibility and the British Wind Energy Association. For him, fair trade purchasing follows from these commitments:

> JOHN: Well I suppose, on one level, you say, I have a very personal obligation so I buy fair trade and that's like discharging some sort of personal obligation or commitment, and then you say, wearing the political hat, and writing letters to MPs or chief executives, you're attempting to influence the wider, have an impact on a much wider scale, and I suppose I'm quite motivated to do that.[32]

While fair trade fits with John's other political commitments, he does not see it as a political act per se:

> JOHN: I mean obviously it would be hypocritical if you didn't do those things yourself, um, but in a way I think one letter that hits home and leads to some small shift in a big organization has a much bigger impact than my own personal consumption of coffee.[33]

This range of personal and public activities is typical of the participants in fair trade networks in and around Bristol. Jennie runs a Traidcraft stall at her church and another stall at the local farmers' market in North Bristol: 'We're actually out in the street and this is what we wanted to do, to try to get it beyond the church doors'.[34] But this is just an extension of other commitments, including her active involvement in trade justice campaigns through her involvement with Christian Aid:

> JENNIE: Well I get the information from Christian Aid. You know, there's sort of an ongoing campaign of sending postcards and such like. And I get the postcards for church, pass on the information, set up displays and such like. And an awful lot of that is to do with trade justice because, as you know, an awful lot of charities are involved with that.[35]

Similarly, Pauline is the representative for Christian Aid at her own church, which is, as she puts it, 'all tied up with trade justice, and fair trade, Drop the Debt, that sort of stuff'.[36] She sees herself as a campaigner rather than an activist. She has attended protest marches at the G8 summit in Genoa in 2000, the Labour Party conference in Brighton in 2004, and the G8 summit near Edinburgh in 2005.

John, Jennie and Pauline are all involved in a variety of forms of individualized activism, of which their consumption of fair trade products is just one aspect. But it is important to recognize that, for each of them, these fair trade activities sit alongside forms of contact and collective activism such writing letters and sending postcards, or going on marches. Furthermore, fair trade is, for them, a supplement to other involvements in trade justice and global poverty campaigns. It serves as a way of extending and performing these public commitments into and throughout their everyday lives.

The 'career' of Edna, one of Traidcraft's Key Contacts in Bristol, embodies these different forms of activism and their articulation with different organized modes of public action. Edna has worked with Traidcraft for 25 years, starting with a personal commitment that quickly spread into her church network:

EDNA: I started buying coffee just for myself, and fairly quickly became more and more convinced that this was the right thing to do and that other people ought to do it too, and started talking to people in church. And we started buying it as a group, and then it just gradually grew from there.[37]

Edna soon became a Key Contact for her area of Bristol, helping other Fairtraders with their stocks and events, and more generally encouraging and advising Fairtraders. But she is increasingly involved with moving fair trade 'out of the church'. This partly involves 'weaning certain people off buying their Cafédirect from me, when I want them to buy it in the supermarkets'. She is also sensitive to the danger that fair trade is seen by some churchgoers as a form of paternalistic charity given to passive recipients in the Third World. Moving out of the church therefore also involves foregrounding the political dimensions of fair trade, and she has become increasingly involved in campaigning, having become actively involved in the Bristol Trade Justice Group, and in Christian Aid campaigns aimed at G8 meetings in Genoa and Edinburgh. She now sees herself primarily as a campaigner rather than just a Fairtrader, reflecting her own gradual realization that, on its own, buying and selling fair trade products is not 'going to change the world': 'The world is actually going to have to change by political will.'[38] Edna's progression – from Fairtrader in church, to Traidcraft Key Contact, to campaigner and activist – epitomizes the organizational diversification of Traidcraft across different functions (trading, charity, lobbying and advocacy) and different social and institutional networks. Edna embodies a series of different roles – customer, seller, activist – that are articulated together in deployment of fair trade consumption as one aspect of the broader fair trade movement.

Edna is one example of the combination of fair trade consumption with various other forms of citizenly acts. Another example is provided by Jennie, whose experience running Traidcraft stalls in her local church for over 12 years provided the basis for a movement into what we might call 'Fairtrade entrepreneurialism' as an activist-retailer (cf. Wempe 2005). Accounts of women's entrepreneurialism (Hanson 2003) emphasize how the gendered geographies of everyday life are intertwined with start-up processes, location decisions and the relationship of the business to place. In Jennie's case, her decision to set up a family oriented, child-friendly fair trade café, Café Unlimited, in Bristol in 2004 was triggered in part by her own experiences as a mother. She had been working in the social housing sector, but was 'finding it more and more difficult trying to juggle work with family', she decided her dream of running a fair trade coffee shop should not wait until her retirement. Jennie recognized that her local area, the Bishopston and St Andrews area of Bristol, was already supportive of fair trade and had a 'community feel' and decided 'if I'm going to do it, this would be the right place'. Jennie's experience of her local community informed her perception of the market opportunity for a fair trade and family friendly café. And her involvement in local church networks already active in trade justice and fair trade provided a dependable customer base, again especially in the crucial early phases. Her 12 years as a sales representative for Traidcraft provided her with a set of established relationships with suppliers of fair trade goods, which she utilized to negotiate larger stock on credit during the difficult early phases of her new business venture. For Jennie, the café, served a dual purpose: geared to families, those with small children with space for buggies, it provided a well-resourced children's play area, weekly free storytelling sessions for children and a relaxed attitude and child friendly policies such as serving the children first, and providing healthy, small portions. It quickly became a regular meeting place for social gatherings, such as postnatal coffee mornings. At the same time, all the drinks were fair trade and, where possible, all cakes and meals made with fair trade ingredients. This business venture thus provided one 'Fairtrade space' in which ethical consumption was configured with rhythms of other practices, in this case of childcare and the routines of sociability surrounding it.[39]

In this section, we have used the reflexive understandings of Fairtraders themselves to further displace 'the consumer' from the centre of analytical attention when seeking to understand the dynamics of fair trade consumption. From their own perspective, taking part in fair trade consumption provides committed Fairtraders and fair trade customer-supporters (who occupy positions of relative privilege by virtue of their location in advanced industrial economies in the global North) with a means of engaging in shared projects which avoid the paternalism popularly associated with charitable appeals and international aid: 'It's a working partnership with other people to try to improve the conditions of those people who really are struggling.'[40] Thus, displacing 'the consumer' helps to bring into focus the ways in which the fair trade movement practically experiments with the challenges of developing a distributed, shared political global responsibility of the sort we discussed in Chapter 1.

6.5 Conclusion

We have argued that fair trade practices in the UK can be analysed with reference to the forms of civic participation and collective action that they help to articulate. Rather than conceptualizing fair trade through the lens of consumer demand, we have focused on its political rationalities. We have used the case study of Traidcraft to elaborate on the relationship between organizational strategy and social networks through which fair trade consumption works. Traidcraft is not simply a trading company operating in the marketplace. It is embedded in a variety of social networks through which it recruits individuals into more sociable and collective forms of civic and political practice. And it operates as a trading company, charitable fund-raiser, and lobbying and advocacy organization, either in its own right or by pooling its resources in networks, coalitions and campaigns.

The prevalent economistic framing of fair trade, in business and management studies as well as in critical social science, reproduces a picture of the world whose key agents are producers and consumers acting through the market. The case study of Traidcraft networks indicates that agency needs to be located not in the activities of consumers, but in the articulation of intermediary organizations, social networks and everyday practices of social reproduction. The political rationality of fair trade organizations does not imagine the subjects of fair trade consumption as individualistic, rational consumers. Traidcraft thinks of its activities in one sphere as supplementing and supporting those in other spheres: it seeks to enrol supporters not as consumers located at one end of commodity chains, but as members of social networks that extend into all sorts of ordinary, everyday spaces. Likewise, Fairtraders and fair trade customer-supporters think of their own fair trade activism as drawing on and extending their own social networks, and they understand fair trade to sit alongside other commitments and energies they already practise.

In both its organizational complexity and in the practices of Fairtraders and their customer-supporters, fair trade consumption combines aspects of individualistic, contact, and collective activism that cross-divides between private life and public sphere, the ethical and the political. Understood in this way, the growth of fair trade consumption in the UK carries important theoretical lessons for accounts of civic and political participation. First, it suggests that these activities can take place in all sorts of mundane locations (at coffee mornings, in school assemblies, at church stalls) at the same time as belonging to spatially and temporally extended networks of advocacy, campaigning and mobilization. Second, it suggests that the key axis of social differentiation around which fair trade consumption is organized is not simply income level, but a more complex assemblage of professional expertise, associational life and social capital in which the key actors are institutions such as churches and schools and trade unions, and in which people participate not as abstracted consumers but as Christians, or socialists, or teachers, or friends. The fair trade movement mobilizes existing, geographically embedded social networks with the purpose of sustaining a vision of alternative economic and political possibilities, networks rooted in local church communities or in localities where local businesses, fair trade activism and willing customers collude to generate thriving fair trade 'scene'. In this chapter, we have explored the non-consumer based identifications that sustain one example of such faith-based, locally embedded networks. In the next chapter, we shift attention to look at the dynamics of block recruitment through which whole communities and networks are mobilized as 'fair trade consumers' – and, in so doing, we examine how the 'mainstreaming' of fair trade involves not only the availability of products on supermarket shelves but also a much a more diverse movement of fair trade as product, practice and meaning into a wide variety of public spaces.

Chapter Seven

Fairtrade Urbanism

7.1 Rethinking the Spatialities of Fair Trade

In the previous chapter, informed by Crossley's (2007) analysis of the role of social networks in sustaining extra-parliamentary politics, we looked at the ways in which social networks and interpersonal ties are an important dimension of the recruitment of individuals into fair trade campaigns. In this chapter, we look at the ways in which these social networks are recruited en bloc, as it were, into campaigns which focus on reconfiguring the material systems of provisioning in which so much consumption is embedded. Analysis of fair trade, ethical trading and sustainable consumption often assumes a relationship involving disparate placeless consumers being stitched together with place-specific producers in developing-world contexts. As we argued in Chapters 3 and 4, conceptualizing consumption as embedded in practices throws the growth of ethical consumption campaigning into a new light. This approach underlines the extent to which this phenomenon is the outcome of the concerted efforts of organizations and campaign groups adopting particular political strategies and repertoires of mobilization. It also suggests that campaigning is likely to succeed not by addressing people as placeless consumers, but by acknowledging their attachments to particular communities of practice.

One example of campaigning that revolves around the block recruitment of social networks is the extension of certification by the Fairtrade Foundation (FTF) beyond consumer goods to various institutional actors such as schools (see Pykett *et al.* 2010), churches, universities, and even places. In this chapter, we focus in particular on the dynamics of the 'Fairtrade Town' device and its extension to large metropolitan urban centres.

Globalizing Responsibility: The Political Rationalities of Ethical Consumption by Clive Barnett, Paul Cloke, Nick Clarke and Alice Malpass   © 2011 Clive Barnett, Paul Cloke, Nick Clarke and Alice Malpass

This device was adopted as a strategy by the Fairtrade Foundation, the leading NGO in the UK committed to campaigning for and regulating fair trade, after a local initiative by a group of Oxfam activists in Garstang, Lancashire, announced their town to be the first 'Fairtrade Town' in Britain in 2000. By the end of 2009, 435 towns and cities had been awarded Fairtrade Town or City certification.[1]

In 2005, Bristol became one of the first large metropolitan authorities in the UK to gain certification as a Fairtrade City. In this chapter, we draw on a year-long ethnographic study of the policy making and political processes involved in the Bristol Fairtrade City campaign from its launch in 2004 to its culmination in 2005.[2] We argue that the Bristol Fairtrade City campaign stands as a practical experiment in developing what Massey (2007) has called 'the politics of place beyond place', in which the political obligations of public authorities, private companies and citizens located in particular places are reframed to include considerations of the relationships of dependence and interdependence that reach far beyond the confines of those particular locations.

There are two lessons we draw from the Bristol Fairtrade City campaign in this chapter. The first is that this campaign was used by campaigners and local authority actors to enlist ordinary people in Bristol, to raise awareness about fair trade issues and support for the city's own bid for certification. This aspect of the campaign depended on the careful calibration of a set of local concerns – finessing Bristol's ambivalent heritage as a trading city, for example – with the more global concerns of the fair trade movement around global trade and the alleviation of poverty. Arguing that the certification would be 'good for Bristol' was a means by which various different interests – including businesses, local government departments, local charities, community groups, trade unions, as well as ordinary people – were gathered together into support for a campaign whose primary beneficiaries are, in principle, people living a long way away, in the Caribbean, West Africa, or Central America.

However, and this is the second lesson we draw from this case study, a large part of the campaign did not actually involve this sort of public campaigning to raise awareness and support at all. Instead, it focused on making the procurement practices of local authority departments consistent with principles of fair trade and sustainability. Here the campaign involved enlisting the support of key professional actors not with the aim of addressing consumers directly, but with the aim of changing the systems of collective provisioning of whole organizations, both public bodies like council canteens and restaurants, as well as important local businesses, like Bristol Zoo and Wessex Water. The Bristol Fairtrade City campaign indicates that campaign organizations operate at different levels to enlist support and transform consumption practices: sometimes they deploy devices that are presented as extending choices to consumers to raise awareness amongst a broad general public and generate media attention; sometimes they engage at an institutional level to change the ways in which consumption is regulated at the level of whole systems of provisioning.

The Bristol Fairtrade City campaign illustrates more than simply a story of local political mobilization, interesting and significant though that is. The campaign offers insights into two often neglected aspects of fair trade activism. First, it brings into focus the emphasis on the collective rather than individual espousal of fair trade; second, it shows the importance of 'emplaced' rather than seemingly 'placeless' consumption of fair trade ideas and goods. Studies of fair trade movements and networks have positioned fair trade as a simple linear linkage between needy and exploited producer communities (e.g., Rice 2001) and individual consumers who are drawn into the commodification of moral betterment (e.g., Goodman 2004; Gould 2003). Fair trade marketing often addresses consumers through politically inflected narratives in which producer communities are explicitly located in particular places (e.g., Bryant and Goodman 2004). These places are often represented photographically, typically showing farmers in-place as part of product packaging and advertising. In contrast, and as we have seen in previous chapters, for strategic reasons, fair trade campaigning in the global North tends to represent participants in fair trade consumption practices as *consumers*.

This representational economy of *emplaced producers* and *placeless consumers* is, in turn, reiterated in much of the academic analysis and evaluation of fair trade consumption. The widespread presentation of fair trade as reconnecting placeless consumers with in-place producers (see Holloway and Kneafsey 2000; Hughes 2001) is predicated on theoretical assumptions about how 'space hides consequences' and 'distance leads to indifference'. Using these assumptions as the starting point for analysis risks losing sight of how 'caring at a distance' often works by instigating the re-examination of responsibilities closer to home (Barnett *et al.* 2005). The analyses of the dynamics of the Bristol Fairtrade City campaign which we develop in this chapter suggests that fair trade 'consumers' should be conceptualized as thoroughly emplaced. Acknowledging this aspect of fair trade consumption allows us to understand the ways in which the 'place-based' ethical consumption campaigning works by drawing together local social networks, of the sort discussed in the previous chapter, with locally dependent public and private enterprises and locally scaled state agencies. A key objective of campaign devices like Fairtrade Towns and Cities, or Fairtrade Schools and Fairtrade Churches, is to govern collective systems of provisioning by mobilizing the collective authority of local authorities, boards of school governors, and the like. The role of local authorities in the re-imagination of place identity in the campaigns such as that in Bristol is also significant in widening fair trade constituencies within local social networks. As we saw in Chapter 6, involvement in the fair trade movement has often been nurtured amongst social networks revolving around faith communities such as churches (see also Cloke *et al.* 2010). The Fairtrade City campaign in Bristol embraced faith-based activists, but the key coordinating role of the local authority ensured a broad disassociation of fair trade from any specific religious identity. Individuals not in faith-based networks can be called on as residents of place to embrace re-imagined place identity without having to associate themselves with religion, a move which is fundamental to the mainstreaming of fair trade beyond niche consumption (and often 'religious') social networks. Local authority jurisdiction opens up the possibility of converting community-based units – schools, libraries, swimming pools, and even in Bristol's case the zoological gardens – as fair trade zones which 'belong' to the city and its people rather than to particular restricted membership (as in churches). Thus the civic re-imagination of place takes root in particular facilities where fair trade procurement policies ensure that fair trade consumption takes place whether or not workers, visitors or consumers consciously choose to participate, or even realize whether they are participating, in the 'urbanization of fair trade' in their city.

In this chapter, then, we analyse how fair trade campaigning uses place-sensitive strategies to develop the type of shared, distributed sense of responsibility discussed in Chapter 1. We focus here on how place is used as a mobilizing device for collective action (Escobar 2001; Miller 2000). In this case, place is mobilized to articulate local state agencies and other locally dependent actors implicated in governing systems of provisioning, consumption practices and trading networks around a shared campaign to achieve certification from a non-governmental body, the FTF (Freidberg 2003; Hughes 2001; Marsden 2000). In pursing this analysis, we develop in empirical detail Massey's (2007) account of the political challenges of organizing 'a politics of place beyond place'. These challenges involve not only acceptance of the openness and relational construction of place so as to develop external relations of identity characterized by an overcoming of self-interest and an acceptance of responsibility, but also the internal political construction of the identity of place so as to bring about a re-imagination of place from the perspective of looking from the inside out. Massey argues that not only can the 'outside' of a place be found within, but also that part of the 'inside' of a place lies beyond, such that other places are in some senses part of our place, and vice versa. This formulation of the 'politics of place beyond place' offers insight into new forms of solidarity between places. Accounts of transnational activism call attention to the crucial mediating role of local networks in anchoring and translating global justice imperatives in and through grassroots action (see Della Porta and Tarrow 2005; Stark et al. 2006). In this chapter, we look at one example of this type of distributed politics of locality in which actions previously constituted as 'aid' or 'charity' are transformed into political acts of obligation.

As a city with its own deep implication in the history of slavery, a local politics of race and racism, and located within a largely rural agricultural hinterland, re-imagining Bristol as a global 'fair trade' place involved the negotiation of different interests, imperatives and imaginaries. By re-imagining Bristol 'from slave trade to fair trade'[3], the Bristol Fairtrade City campaign was one moment in a longer-term re-imagination of the city's place in the world. This effort first came to the fore in the early 2000s, when the local council led an unsuccessful bid to the European Union's European City of Culture programme. This re-imaging of Bristol has continued through various initiatives, such as the 2007–2008 city-wide campaign to mark and com-memorate the bicentenary of the Atlantic slave trade. This set of campaigns draws on and is sustained by an extensive network of civic organizations in Bristol active around global and environmental issues (see Diani 2005; Purdue *et al*. 2004).

A particular alignment of interests led to a diverse group of individuals, networks and organizations in and around Bristol adopting the Fairtrade City device. The outcome of the campaign was not only to engage residents of the city with the aims and agendas of the fair trade movement; it also led to significant changes to public procurement practices, bringing about new models of collective provisioning (cf. Morgan and Sonnino 2007; Sonnino 2009). The campaign therefore illustrates the extension of the politicization of consumption beyond individualized consumer choices into novel modes of collective action, not least with the aim of leveraging the jurisdictional enforcement of fair trade provisions with or without the formal consent of individuals embedded in consumption practices. In this way, the Fairtrade City story illustrates how place can become a territory for a configurational politics in which individuals are joined into wider communities of action.

7.2   Re-imagining Bristol: From Slave Trade to Fair Trade

The Fairtrade Foundation's desire to develop the initial Garstang campaign into a nationwide programme was given warm central government approval in 2002, with specific endorsement from the Department for International Development and its then under secretary, George Foulkes MP, who told the FTF to 'ensure that the initiative is followed in many other towns and cities throughout the United Kingdom and beyond'.[4] Accordingly, the Fairtrade City campaign was designed to provoke the collective imagination of place and community in particular towns and cities. It also aimed to fuel rivalries between places which could be exploited to ensure widespread par-ticipation in the programme. The aim of the campaign highlighted the need to 'tackle poverty by enabling disadvantaged producers from poor countries to receive a better deal through encouraging support for the Fairtrade mark'.[5] The specific objectives of this device for the FTF were to develop national and regional support for the fair trade mark, to increase levels of awareness and understanding of the fair trade concept, to increase the avail-ability of fair trade products and to increase sales of fair trade products.

These aims and objectives translate into five criteria which candidates for fair trade cities and towns have to attain:

1   The local authority must pass a resolution supporting fair trade and agree to serve fair trade tea and coffee at its meetings, offices and canteens.
2   A range of fair trade products must be made available in a specified number of local shops and cafés, so that it should be easy for local people to include fair trade products in their everyday shopping.
3   Fair trade products must be used in a specified number of local work-places and community organizations.
4   Significant media coverage and popular support must be attracted for the campaign, and a strategy must be developed to keep the campaign in the news.
5   A local fair trade steering group must be established to ensure continued commitment to the Fairtrade City status.

To these mandatory elements, the FTF add further options, including: reg-ular local authority promotion of fair trade awareness; local authority allo-cation of Fairtrade City responsibilities to a particular staff member (often Environmental or Agenda 21 Officers); erection of street signs declaring the place as a Fairtrade Town/City; production of a local fair trade directory; inclusion of a flagship employer in the matrix of support; and organization of workplace educational campaigns and educational awareness events or competitions.

The Fairtrade City device is therefore envisaged as a way of focusing and widening existing fair trade activity. The FTF recommends that local Fairtrade City steering groups should include local authority representa-tives alongside campaigners from businesses, churches and schools. The incorporation of the local authority as integral to the campaign strategy means that the formal authority of political jurisdiction is combined with the voluntary engagements of companies, social networks and individuals. Successful collaboration can be stymied if there is no initial impetus from the local authority. In the Bristol case, a Fairtrade steering group was quickly established in 2003, energized by an ex-employee of the FTF who was then working for the local authority's sustainability department. This person became a key actor in the campaign. The significance of her role was magnified by the coincidental appointment of two other key local authority officers at the same time – a new manager of the Corporate Procurement Unit (CPU) who was also appointed as Fairtrade officer for the council (a job specifica-tion created for the first time in the local authority) and a sustainability officer working both on food policy and procurement, who was a long-term personal supporter of fair trade. These serendipitous appointments were a key situational factor behind the development of the Bristol Fairtrade City campaign.

Beginning with these three local authority professionals, Bristol's Fairtrade City campaign steering group quickly brought together a range of other activists and interests from around the city. These included faith-based activists from Christian Aid, Traidcraft and Oxfam; ethical business interests representing local food retail outlets; Equop, a local organic fair trade clothing company; locally based national interests such as the Triodos Bank and the Co-operative; and representatives from school and university groups. Trade union representation was not part of this initial configuration of interests, an absence which has been common in Fairtrade Town and City campaigns across the country. Within the existing coalitions of interest representing sustainability, faith, social solidarity and poverty reduction, there are significant differences in terms of how widely or narrowly 'fair trade' is defined. The specificities of how the Fairtrade accreditation campaign 'gets into place' enable particular balances between maximizing activism and minimizing potentially damaging divisions of interest. In particular, the re-imagination of particular places is crucial to forming and sustaining these coalitions of interests. In the Bristol example, the emphasis in kick-starting the Fairtrade City campaign was on a strict timeline within which to achieve Fairtrade accreditation. A key objective was keeping the campaign simple in order to rise above potentially conflictual local interests, and to present a 'win–win' outcome to all concerned. For this reason, the Fairtrade City campaign was not initially aligned with green interests that are also an important presence in the political culture of the city. But when potential conflict arose between fair trade and issues of fairness to local producers, these local fair trading objectives were integrated into the overall campaign.

The reframing of place in the Bristol Fairtrade City campaign involved a practical working through of the politics of scale. In part, negotiating the potentially competing interests which have been mobilized under the Fairtrade City banner can be understood in terms of the deployment of 'scale frames', discursive practices that construct meaningful linkages between the spatial scale at which a matter of concern is experienced and made relevant and the spatial scale at which action needs to be taken to effectively address this issue (Kurtz 2003). We look at this deployment of scale frames in the Bristol Fairtrade City campaign in more detail in the next section. Here, we want to underscore the degree to which the Bristol Fairtrade City campaign involved a politics of scale in the sense that it involved locally dependent actors aligning their interests while also drawing on spatially extensive networks (Cox 1998). The ways in which key actors in the Fairtrade City campaign were embedded in particular scalar networks provided crucial impetus for their involvement in the Fairtrade City campaign.

One important dimension of this scalar politics of the Fairtrade City campaign is the relationship between local authority responsibilities and 'sustainable development' targets defined by central government. In particular, since 2003 central government policy around 'sustainable procurement' helped to shape the role played by local authority officers in the Bristol Fairtrade City coalition. These policies, produced and published by the Office of Government Commerce (OGC), have effectively redefined 'value for money' in public procurement as meaning more than 'least cost', acknowledging that 'value for money is not an end in itself'. This acknowledgment is reflected in a series of policy statements setting out procurement principles in relation to environmental, green and social issues. In 2004, the OGC issued policy guidance to all Whitehall departments to coincide with the start of Fairtrade Fortnight, the FTF's annual campaign to maximize and coordinate all of its regional activities relating to fair trade. These guidelines clarified that 'best value' conventions for the public procurement of goods, works and services need not be seen as counter to fair trade connections of a 'fair price':

> Value is not just about price. It is defined as the optimum combination of whole-life cost and quality to meet the user's requirement.[6]

As we discussed in Chapter 2, these sorts of central government policy agendas operate through the active interpretation of rules by locally emplaced professional actors (Morgan and Morley 2004). These cues were rapidly acted on within Bristol City Council. In the Bristol case, 'value for money' and sustainability criteria were creatively interpreted to make space for fair trade procurement initiatives. A new sustainable code of practice was established which sought both to establish forms of local governance with which to dislodge existing conventions of 'best value', and to remove any hurdles from EU directives which might otherwise prevent the introduction of fair trade issues into broader sustainable procurement practices. The Council's Procurement Code of Practice concludes that they should attempt to govern production conditions elsewhere by including their fair trade and sustainability requirements at the point of specification and tendering. In this way fair trade specifications became enfolded into the everyday procurement practices of the council.

These changes were facilitated by the formation by the local authority of a new Corporate Procurement Unit in 2003 responsible both for streamlining procurement practice and introducing fair trade and other ethical specifications. These potentially strange bedfellows of policy change are no coincidence, since the local authority officer charged with overseeing the rationalization of procurement was also the council's sole appointed Fairtrade officer. In combination, these roles present the most effective channel for rebuffing potential financial and legal challenges to fair trade as a significant new purchasing convention. The Corporate Procurement Unit's challenge was to reduce and rationalize the existing number of suppliers. At the time Bristol City Council had 19,374 different suppliers of contracts. Fair trade has thus become aligned with a 'move from existing arrangements into a more corporate perspective'.[7] Just as in wider FTF campaigns, the Bristol Local Authority initiative was seeking to shorten the producer–consumer link, and so the aggregation of contract suppliers was seen to fit a move towards fair trade, not least because both streamlining suppliers and introducing fair trade principles was about managing business relationships differently, 'to make sure you keep a good business relationship with a reduced number of suppliers'.[8] It was about making the supplier 'known', building the business relationship, restoring to view a previously hidden set of complex relationships.

These procedural changes beg the question of what exactly the Fairtrade City campaign achieves for local authorities. Why should a city such as Bristol work so hard simply to get a certificate from a national NGO? In this case, the Fairtrade City idea was an ideal vehicle for bringing together different responsibilities, not least centrally directed requirements for 'sustainable' and aggregated procurement, and local sustainability initiatives such as the Bristol Community Strategy launched in 2001. The place-based jurisdiction of the local authority permits amplification and extension of the framing of different though not necessarily conflicting interests. The Fairtrade City idea represents a politically neutral way to support a campaign aiming to reduce distant poverty whilst keeping to a repertoire of local jurisdiction. In this way the Fairtrade City provides a repertoire of political agreement and helps to build support within and between communities and across different political interests. The Fairtrade City campaign therefore relies on particular discursive strategies through which actors can 'speak for the city', thereby negotiating the potentially competing interests which have been mobilized under the Fairtrade City banner. Thinking of these discourses as 'scale frames' (Kurtz 2003) helps to explain how campaigns are able to align different interests around shared focal points. Scale-specific storylines can be generated which gather together diverse interests in order to frame the shared recognition of and responsibility for trade injustice. In the case of the Bristol Fairtrade City campaign, scale frames and associated storylines succeeded in operationalizing a consensual space which helped to re-imagine 'place beyond place'.

7.3   Putting Fair Trade in Place

Bristol's Fairtrade City Steering Committee quickly realized that the consumption of fair trade products should not be presented just a way of 'voting with your wallet'. Rather, it offered an opportunity to enrol people in the fight to eradicate poverty through better trade terms. Thus one of the storylines in the campaign emphasized that consuming fair trade products through the Fairtrade City device was an integral part of wider practices of good local citizenship and place-belonging, of becoming

what Escobar (2001) calls a 'placeling'. Fairtrade City campaign literature frames involvement around place specificity:

> Bristol is a great place to live, where we pride ourselves on our maritime his-
> tory and our position as an economic and cultural centre for the south-west
> [. . .] Despite Bristol's trading past, its future is that of an internationalist fair-
> trading city.[9]

This sort of promotion of the city's distinctive qualities as a place are key to
Fairtrade City campaigning, which interconnects the conspicuous consumption of fairness with the endowment of the city with meaningfulness
and civic pride. Thus a Fairtrade City becomes not only a place that is
known to promote fair trade but also a place characterized by fairness. As a
member of the Bristol Fairtrade City steering group describes:

> People think 'Oh the council is not exploiting people *outside* the city' so per-
> haps people might also extrapolate that the council is not trying to exploit
> people *in* the city.[10]

In other words, in some ways the Fairtrade City badge becomes thought of
by local authority representatives, both professionals and elected politicians,
as a way to promote the local reputation of the authority and the way it acts
within its own jurisdiction. Bristol City Council's Code of Practice suggests
that the risks of non-compliance with sustainable and fair trade specifications will bring damage to the council's reputation. Temporal aspects of
these constructions are integral to the evocation of a sense of fair trade,
which becomes not only a distinguishing mark of the newly created fair
trade 'zones' but also a temporally resonating marker that builds on and
reframes existing iconic images of the city and its people. The publicity for
the Fairtrade City campaign quotes a local television news journalist,
Sherrie Eugene, the daughter of first generation Caribbean immigrants,
who grew up in the city:

> Fair trade is about recognizing that everyone has worth. It's about being bal-
> anced, conscious and fair. *Let's make Bristol known* not just for the Suspension
> Bridge or the Slave Trade but also for Fair Trade.[11]

This statement comes from a promotional leaflet that was distributed to
residents of Bristol as part of the year-long campaign in 2004 and 2005 to
achieve Fairtrade City accreditation. Mobilizing iconic images draws not
only upon a positive historic past, such as the industrial triumphs of technology signified in the Clifton Suspension Bridge, and the positive association of fair trade with celebrities in the media, but also upon the negative
iconography of past trading relations, the slave trade upon which Bristol's
wealth and infrastructure was built. Redirecting people's attention to their

own sense of emplacement as historical, situated subjects is used as a way to engage people with fair trade. This strategy seeks to promote an inter-subjective relationality by asking people to look around them, at their build-ings, bridges, celebrities and fellow residents, and to ask 'How did this come to be?'. For those involved in the Fairtrade City campaign, the use of iconic images, eras and people is an understood device to localize fair trade and make it relevant to people's local concerns:

> putting the focus back on Bristol, people tend to think fair trade is something that happens elsewhere [. . .] By talking about the slave trade in Bristol, how those historic links have made a difference and how now we can make this [fair trade] a more positive difference [. . .] I think it's really good to hook people in on something that they do know about and perhaps learnt about at school or read in the newspapers.[12]

The Fairtrade City campaign sought to re-imagine and even redeem specific place-histories of Bristol. In this way, local authority actors – politicians and officials alike – sought to attract for themselves some of the positive features associated with fair trade. Fairtrade City status works for the city as much as the city works for fair trade. By connecting place-imagination to fair trade, the local authority gains a sense of worthiness. By bringing together previously disparate sets of interests, the city gains a sense of unity, with people perhaps beginning to realize that they have something in common with the council. By sponsoring a sense of investment in place, the city gains a sense of commitment, which can be reinforced through Fairtrade City signs which present a symbolic re-evaluation of the cam-paign and its success to residents and visitors. Most obviously, the Fairtrade City idea enrols numbers of people in the sense that the entire city can claim to be participating in the political virtue of fairness. The framing of place through ideas of 'fairness' in the Bristol Fairtrade City campaign, therefore, can be seen as a device to broker deals between different interest groups. This brokerage involves the enrolling of various parties into the idea of a unifying 'win–win' place-based identity, whilst simultaneously entwin-ing that sense of identity with the process of working up that place as a showcase of moral virtue.

7.4   Fair Trade and 'The Politics of Place Beyond Place'

Massey's (2007) conceptualization of how the 'outside' of a place can be found within and how the 'inside' of a place can be found somewhere beyond questions forms of politics which privilege the 'inside' of any place, and gestures instead towards a vision of a politics of solidarity that works through rather than in spite of place-based affiliations. The Bristol Fairtrade

City campaign deployed spatial imaginaries of local and global in distinct ways. Not only is the emphasis on the overriding objective to improve the position of poor and marginalized producers in the developing world but there is also a clear intention to re-envisage and if necessary 'support' the local.[13] There is potential for conflict between these two emphases. On the one hand, the local authority's championing of fair trade procurement has resulted in the promotion and use of Fairtrade-certified products in local authority canteens and offices, suggesting a preference for global sourcing underpinned by ethical criteria of social justice. On the other hand, there are strong pressures for the local authority to give preference to local suppliers of services and products, suggesting a contraction away from global sourcing. The Fairtrade City campaign negotiated these potential conflicts using a storyline grounded in the idea of the local and global being mutually constitutive. Thus, the locality was characterized by reference to its past and present interconnections with the wider world, thus necessitating a global sense of the local. At the same time, global networks of trade, travel and communication are presented as being routed through this place, necessitating a local sense of the global. Alongside these mutually constituted local–global relations, the Fairtrade City campaign also emphasized the idea of 'making the local link'. Local links have taken two forms. First, there has been a commitment to advertise, promote and procure 'local' produce as part of espousal of 'fair' trade. Secondly, there have been significant attempts to arrange exchange visits through which connectivity can be achieved between local producers and consumers and fair trade producers in the developing world. The original emphasis of the FTF, on 'far off' producers had to be speedily reviewed in the light of the experience of the first Fairtrade Town campaign in Garstang, as explained by the coordinator of that campaign:

> When I saw dairy farmers marching down Garstang High Street carrying a banner bearing the words 'we want a fair share of the bottle', I realized that we could no longer continue campaigning on fair trade with developing countries, without making the link to local farmers. They also want a fair share for their produce.[14]

The Garstang campaign agreed to a political alliance with local dairy farmers, and this even opened up a much wider critical debate concerning the tension and conflicts of interest between fair trade as international trade justice and fair trade as a more generic fair deal for local producers. Such debates have been extensively reviewed in literatures relating to the localism of food in which claims relating to the justice and sustainability of local support for local farmers have been pitched against counter-claims about defensive localism which acts to obscure the more significant injustices of international trade (e.g., Hinrichs 2003; Winter 2003). Similar discussions can be found in the Fairtrade Town online discussion forum. For example, Bruce Crowther, a member of the original Garstang campaign admits:

> We were aware of this conflict from the start which is why we took so long in making the connection. The connection we made, and felt we could not avoid any longer, was a simplistic one, i.e. that farmers in the UK also want, and indeed deserve, a fair price for their produce.[15]

While many campaigners believe that 'the ethics governing local produce and organic production are true also for fair trade'[16], others are more sceptical of confusing similar, but quite different, types of ethical trade issues. For example, while price control by large corporations and the habitual undercutting of small producers by large farms is seen as a common problem in both the UK and overseas, the wider campaigns from Oxfam and Christian Aid to 'Make Trade Fair' by removing protectionist legislation have provoked distinct conflicts of interest between local and global producers. In the event, most Fairtrade Town and City campaigners have found it difficult to gain local momentum for their campaign without harnessing the support (and thereby avoiding the opposition) of local agricultural producers. Fairtrade Town and City campaigns therefore couple global and local fair trade, using the scale frame of place to bring diverse interests together.

In the Bristol Fairtrade City campaign, considerable debate took place over whether alignment between 'local' and 'global' producers was advisable under the banner of fair trade:

> If you were looking at UK farms, it's a bundle of different issues basically . . . . If you are going to be accountable, you have got to be very clear about what those standards are and very confident that you can monitor those standards; it's really important to link the issues but it's important not to confuse the Fairtrade mark with other things . . . there's a lot of incomprehension around still about exactly what fair trade means.[17]

> It depends how strongly defined fair trade is I suppose. If it's fair trade as in the Fairtrade Foundation logo then yes it is just tea and coffee but if you are talking about trading more fairly with local people, then it becomes a far more complicated issue . . . we mean just ethically trading, trying to promote the wellbeing of the city rather than going out to corporate giants all the time in a way.[18]

The latter viewpoint, grounded in the sustainability agendas which were part of the initial bedrock of the Bristol bid, prevailed in the Bristol case. 'Local' and 'global' understandings of fair trade were bundled together in the Fairtrade City campaign.

'Making the local link' has also involved the formation of partnerships between Fairtrade Cities and fair trade producers in the developing world.

For example, after Garstang became the first Fairtrade Town, it set up a twinning link with a fair trade cocoa farming community in New Koforidua, Ghana. The link enables 'community exchanges' between local dairy farmers in Garstang and cocoa farmers in Ghana, with the aim of 'finally helping to break down the barrier between local farmers and those in Ghana enabling a closer relationship to develop to the benefit of both communities'.[19] This repertoire of agreement has used a local–global scale frame to enrol two different interest groups under a common banner.

These sorts of international linkages are common in other Fairtrade City and Town campaigns, though with different spatial imaginations of the local–global relation underpinning them. The Bristol Fairtrade City campaign, with its nascent partnership with fair trade banana producers in the Caribbean, tends to frame fair trade using vocabulary which is both local and global, suggesting a mutually constitutive process:

> It is getting across that fair trade is also local as well as global . . . in that on the one hand it's enlightened self-interest, it's a well-known term in these kind of worlds, in that what you are doing may not obviously give you a benefit immediately but by supporting growers on the other side of the world you are helping them to create more sustainable livelihoods as well, which is to the benefit of the planet as a whole, which is seeing the local within its global context. What other benefits does it have? Well it does also have the feel-good, the motivational benefits for people on this side of the planet as well, so that Bristol can be perceived as somewhere where fair trade is important to the people in the city.[20]

Such a framing of fair trade suggests more than a simple stitching together of UK consumers and overseas producers. In this case, both consumers and producers are placed, and while the Fairtrade City campaign aims to increase sales and awareness of goods bearing the Fairtrade mark, this task has inevitably been entangled with the need to rethink and politicize 'fairness' closer to home. The Fairtrade City campaign has therefore had the effects of raising awareness about fairness in its broader context.

To some extent, the 'bundling' of intra-local and extra-local versions of 'fair' trade can be interpreted as a kind of political expediency which is necessary to diffuse the 'over-there-ness' or even religiosity of ethical consumption. However, one notable outcome of many Fairtrade Town and City campaigns has been the raising of the profile of fairness within the web of relations in which the participating cities are imagined. Fairtrade Town and City campaigning prizes open a more relational construction of place, developing identity around external relations of responsibility and justice, but it has also had an impact on aspects of the internal political construction of place, raising issues of fairness which transcend any local–global divide. Thus as Bristol is re-imagined as a Fairtrade City, the outside relations of place are rendered visible in signs and in particular zones of fair trade activity within the jurisdiction and territory of the city. In a more shadowy and undeveloped way, the development of twinning schemes, and the embracing of concepts of 'fairness' have begun to open up new possibilities for recognizing part of the identity of the city that lies beyond its place-boundaries. To be a placeling of Bristol is to be entrained, knowingly or not, willingly or not, into the jurisdictional strategy and practices of fair trade promotion and provision. Participation comes through membership of and identification with the place jurisdiction of the local authority. By recruiting the local authority actors, the fair trade movement embeds fair trade into place-identity and placed systems of governance over local provisioning. Simply to live in and identify with such jurisdiction enrols ordinary local people into a process which re-imagines their place both from the inside looking out, and from the outside looking in.

7.5 Conclusion

We started this chapter with the observation that critical understandings of fair trade, ethical trading and environmental sustainability often portray placeless Western consumers being stitched together with place-specific producers in developing-world contexts. These producers will often be personified in packaging and advertising, whereas consumers are faceless and placeless, an amorphous group somehow susceptible to the commodification of moral betterment. The result is a sense that the only spatial relation of relevance in understanding these initiatives is 'distance'. This idea of the placeless consumer is grossly oversimplified. In Chapter 6 we showed that fair trade consumers will often belong to particular networks, often associated with particular sites (ranging from churches to Oxfam shops to school-based events); and they will gravitate to particular hotspots of ethical consumption activity, such as the 'shabby chic' cluster of fair trade and organic cafés and pubs, charity shops, and alternative food shops along the Gloucester Road in the area of Bishopston in Bristol – described as the 'last great British high street' by one national newspaper.[21] Fair trade consumption, in short, has never been placeless.

In this chapter, we have focused on a specific form of place-based fair trade consumption. The case study of the Bristol Fairtrade City campaign provided in this chapter indicates two dimensions of what we suggest might be an emergent mode of 'Fairtrade urbanism'. First, the Fairtrade City campaign was used by key actors to enlist people in Bristol, to raise awareness about fair trade issues and support for the city's own bid for certification. This aspect of the campaign depended on the careful calibration of a set of local concerns – for example playing on Bristol's heritage as a trading city – with the more global concerns of the fair trade movement around global trade and the alleviation of poverty. Arguing that the accreditation would be 'good for Bristol' was a means by which various different interests, including business, local government departments, local charities, community groups, trade unions, as well as ordinary people, were gathered together into support for a campaign whose primary beneficiaries are, in principle, people living a long way away, in the Caribbean, or in West Africa, or in Central America. The Fairtrade City campaign thereby enrolled residents into a re-imagination of the city which drew identifications with this particular place into more extensive relationships with other places – by adopting a vocabulary and imagery associated with fairness and justice, the Fairtrade City campaign in Bristol has been an important moment in the process by which local identities have begun to be reframed in terms of relational connections with various 'elsewheres'.

Second, the Fairtrade City campaign was used to enlist the jurisdictional authority of local government, not least through the making of procurement practices within the city council consistent with principles of fair trade and sustainability. Such a move does not address consumers directly, but aims to change the system of collective provisioning of an entire organization. As one respondent told us, this is a significant culture change:

> I think it is more a culture really, a mind shifting, to see that it [fair trade] is important. Obviously once you've written the code of practice you have to make sure it is implemented, that's going to be the difficult thing really. . . I work on individual contracts that come up for re-negotiation or for re-tender and I work again with corporate procurement on individual tenders to make sure that these [fair trade] issues are incorporated. . . . I'm going to give it a while for this to embed down and carry on working with the corporate procurement unit and then suggest they make this code of practice a policy.[22]

Successful changes to procurement policy in the local authority, and in other public and private organizations, not only offer symbolic support to fair trade campaigning, but also ensure that employees, local visitors and tourists from further afield will be consuming fair trade products, knowingly or unknowingly, when visiting the canteens and restaurants of these organizations. Whereas ethical consumption is usually portrayed in terms of conscious consumer choice, transforming procurement policy can go so far as to withdraw choice from the consumer at the point of purchase. These days, parents trailing their toddlers around Bristol Zoo who fancy a hot drink to sustain their spirits will have to 'choose' fairly traded tea or coffee whether they like it or not. Here, then, we can see a transformation of consumption practices prompted at an institutional level through the regulation of consumption through fixed systems of provisioning.

Our account of emergent logics of 'Fairtrade urbanism' in this chapter suggests that local state actors can be enrolled into forms of 'non-governmental politics' to deploy their jurisdiction over place to begin to engage with the political challenges of the place beyond place. There are then two dimensions to the process we have dubbed 'Fairtrade urbanism': educating, informing and engaging residents about shared responsibilities of place that stretch beyond the local; and the use of jurisdictional power to change collective infrastructures of consumption. In the latter of these dimensions, the introduction of fair trade procurement practices in public organizations and private companies alike is the means through which employees, residents and visitors became fair trade 'consumers', knowingly or unknowingly, when visiting the canteens and restaurants of the local authority and other significant organizations in the city. In the former, the ordinarily sociable spaces of the city itself – nurseries, schools, churches, museums – are effectively transformed into surfaces for the public circulation of the images and iconography of fair trade campaigning. And in both of these respects, we see again that this is an example of ethical consumption campaigning in which practical rationalities address people not primarily as 'consumers' but seek to engage with 'thicker' attributes of their personal identifications – as parents, as members of faith communities, or as professionals; and they seek to change not just individualized consumer choices, but to transform systems of collective provisioning 'behind the backs' of consumers by transforming the design and management of infrastructures of consumption.

Chapter Eight

Conclusion: Doing Politics in an Ethical Register

8.1 Beyond the Consumer

In this book, we have focused on the dynamics of political campaigning in which the registers and repertoires of consumerism are used to enrol support for 'global' issues, and used in turn to redistribute this support into various other public realms. We have presented research on the ordinary concerns that people bring to their consumption choices (Chapter 5), as well as research on the role of organizations in expanding the 'ethical' consumer market (Chapters 3 and 6), to argue against widespread assumptions that consumerism is necessarily inimical to the values of citizenship. We have done so, however, without overestimating the citizenly energies that ordinary people can or should be expected to direct to the most mundane features of their everyday lives. We have located the growth of ethical consumption in a wider context of the emergence of new styles of politics, distinguished by the discursive and practical repertoires of mobilization deployed in campaigns (Chapters 4 and 7). Adopting this genealogical approach in order to better understand the dynamics of this phenomenon, we have shown that the primary factor behind the growth of ethical consumption is not the aggregated and anonymous signalling of individual demand in consumer markets, but the motivated mobilization of support by organizations with clear political objectives, which in turn provide pathways for ordinary people to engage in various forms of civic and political participation.

Our analysis has been informed by research that recognizes consuming, exchanging, and shopping as thoroughly social activities (Miller 1998), and thinks of consumption as embedded in practices (Shove 2003; Warde 2005); and research that seeks to understand the politicization of everyday

Globalizing Responsibility: The Political Rationalities of Ethical Consumption by Clive Barnett, Paul Cloke, Nick Clarke and Alice Malpass © 2011 Clive Barnett, Paul Cloke, Nick Clarke and Alice Malpass

consumption by using the tools and concepts of political science and political sociology (Micheletti 2003). Our critical development of these literatures in Chapters 1, 2, 3 and 4, informed by Foucault's notion of studying modes of problematization, led us to focus on two related aspects of ethical consumption: understanding the role of *intermediary actors* in defining everyday consumption as an ethical terrain and redefining consumer choice as a vector of political agency; and understanding the ways in which *ordinary people* engage with attempts to enrol them into broader projects through a register of ethical 'responsibility'. In seeking to understand these two aspects of ethical consumption, we have drawn upon concepts from governmentality theory, conceptualizing the growth of ethical consumption in terms of strategic interventions into everyday practices through the dissemination of discourses and devices that encourage and enable people to adopt new patterns of consumption for a diverse range of reasons.

In putting these theoretical ideas to work in empirical analysis, we have tried to shift attention away from the problem of subject-formation that characterizes so much critical social science research on neoliberalism and advanced liberalism. We have argued that theories of neoliberalism and advanced liberalism pay inadequate attention to the *interactive dynamics* involved in programmes which seek to reshape everyday conduct 'at a distance' through ethical registers of responsibility. This emphasis in both neo-Marxist perspectives and governmentality theory requires revision in light of empirical analysis of the organizational rationalities of ethical consumption campaigns (Chapters 3 and 4), and in light of empirical analysis of the practices of personal and social engagement through which people become involved in ethical consumption practices (Chapters 5, 6 and 7).

Our emphasis has been on elaborating how ethical consumption campaigning extends beyond the usual scenes of consumerism (the high street, the shopping mall, the supermarket) and also addresses a host of actors beyond the much celebrated and much maligned 'consumer'. The emergence and evolution of ethical consumption campaigning illustrates the shared sense amongst diverse campaign organizations that changing the ways in which markets work involves more than simply changing people's consumer behaviour: it requires concerted political action. The contingent deployment of consumerist repertoires needs to be understood in the context of this shared strategic orientation. Ethical consumption campaigns operate at different levels to enlist support and transform consumption practices: sometimes they deploy devices that are presented as extending *choices to consumers*, which help to raise awareness amongst a broad general public; sometimes they engage at an institutional level to change the ways in which consumption is *governed* at the level of whole systems of provisioning. And organizations link up local campaigns with wider international campaigns, in this way serving as mediators for the nascent development of 'global citizenship' (see Desforges 2004).

8.2   Doing Responsibility

A recurrent theme that is raised by the topic of ethical consumption is the claim, or worry, that it reflects a substitution of publicly oriented collective participation by identity-based, individually motivated and privatized forms of concern. We have argued instead that the emergence of ethical consumption is indicative of strategies that seek to provoke new forms of global feeling which are helping to reinvent political participation and civic activism. This claim is based on the argument that consumption needs to be theorized as a facet of integrated and dispersed practices, and that politicizing consumption involves various efforts at problematizing aspects of the practices in which people are entangled. The growth of ethical consumption is often understood in terms of the role of effective consumer demand as the medium through which the ethical preferences of consumers and the ethical records of businesses are signalled in the marketplace. From this perspective, markets are perfectly capable of expressing people's ethical, moral or political preferences just as long as appropriate informational strategies are developed (e.g., marketing, advertising, labelling and branding). In government initiatives on sustainability, in campaigning around the environment, and across the range of 'ethical' trading initiatives, it is often supposed that the main challenge is to provide people with more information in order to raise awareness of the consequences of their everyday consumption choices: then they will magically change their behaviour. However, as saw in Chapter 5, people don't necessarily lack information about fair trade, organic food, environmental sustainability or Third World sweatshops. They actually seem very aware of these types of things, but they often do not necessarily feel that they have the opportunities or resources to be able to engage in alternative consumption activities. Nor do they necessarily feel that they should have to.

We have traced the ways in which people from varied social worlds engage with the multiple demands for them to act responsibly in relation to various global issues. Chapter 5 demonstrated that, by their own accounts, the capacity of citizens to engage with contemporary problematizations of personal and political responsibility is differentiated by their command of material resources – ethical consumption is often considered costly, complex and difficult; and as placing unrealistic demands on people as consumers by ignoring other identifications and obligations. We also showed in Chapter 5 that differently positioned individuals can draw on various forms of cultural capital which enables them to 'answer back' to demands to be 'ethical' and 'act responsibly' in different registers, from frustration and exasperation to criticism and reasoned scepticism. In Chapter 6 we showed that engagement with the problematization of responsibility is also shaped by forms of *associational culture* to which people belong and which shape their capacities to transform embedded practices.

We have also argued that it is important to remember that a great deal of the consumption which people do is not undertaken as 'consumers' at all. Much of it is embedded in practices where they are being parents, caring partners, football fans, good friends and so on. Some consumption is used to sustain these sorts of relationships: giving gifts, buying school lunches, getting hold of this season's new strip. But quite a lot of consumption is done as the background to these activities, embedded in all sorts of infra-structures over which people have little or no direct influence as 'consumers'. Chapter 7, on Fairtrade urbanism, examined a form of ethical consumption campaigning which builds on exactly this sort of acknowledgement. And what both of these points suggest is that the problematization of consumption needs to be understood as a process which addresses people as more than just rational utility-maximizers, because quite a lot of consumption is not sustained by consumers at all.

Thinking of consumption in these terms – as embedded in practices – throws the dynamics of ethical consumption campaigning into a new light. The emergence of ethical consumption is not simply about spontaneous changes in consumer demand being met by more or less elastic market supply. It is an organized field of strategic mobilization and campaigning. Ethical consumption campaigning seeks to embed peoples' existing dis-positions of care, concern and solidarity into the global politics of mobiliza-tion, activism, lobbying and campaigning around issues of trade justice, human rights and environmental sustainability. Sometimes campaigns use products to make contact with ordinary people and to raise awareness of campaigns before enrolling people in more 'active' forms of political engage-ment, like donating, joining as a member or volunteering. Sometimes they use the purchases of 'ethical' products like signatures on a petition, namely as evidence of support and legitimacy for their campaigns and for validation to their own constituencies.

In short, the expenditure of money by people as 'consumers' on ethical products is best thought of as a means to an end rather than the end in itself. It is less an individualistic economic act, and more of a means of acting in relation to larger collective projects which, as we argued in Chapter 1, innovate new forms of shared responsibility. This is an important point to emphasize, because it underscores that, to the degree that the emergence of ethical consumption is indicative of a concerted movement in which dis-persed actors seek to take on a share of responsibility for the consequences of globalized markets, this is not at all equivalent to assuming that this responsibility falls solely on the shoulders of consumers.

Our research on the practices and organized dynamics of ethical consump-tion suggests that the aims and objectives of consumer-oriented activism are best understood in terms of providing people with means of registering their support for particular causes, support which draws on various motivations, rather than seen narrowly as providing a means of directly altering market conditions by exercising purchaser power. And this means that the potentials of ethical consumption might be *less* significant in purely economic terms than is often claimed, in so far as it does not represent a spontaneous expression of consumer demand. It also means that ethical consumption might be *more* significant in political terms than is often acknowledged, in so far as it is an aspect of new forms of organization, campaigning and mobilization around issues of global trade, environmental justice, human rights, world poverty and social justice that stretch the understanding of what is political and how being political can be performed.

Notes

PREFACE AND ACKNOWLEDGEMENTS

1 For further details, see www.esrcsocietytoday.ac.uk/ESRCInfoCentre/View AwardPage.aspx?AwardNumber=RES-143-25-0022-A (accessed 16 January 2010).

2 See www.consume.bbk.ac.uk/ and www.esrcsocietytoday.ac.uk/ESRCInfo Centre/ViewInvestmentDetails.aspx?data=L6KWBgDmJagj5OkyIb6Z%2b%2f rHtgMJio45&xu=0&isAwardHolder=&isProfiled=&AwardHolderID=&Secto r= (accessed 16 January 2010).

CHAPTER ONE: INTRODUCTION: POLITICIZING CONSUMPTION IN AN UNEQUAL WORLD

1 Rebecca Smithers, 'Ethical spending has tripled in 10 years, say Co-op', *The Guardian*, 30 December 2009.

2 Co-operative Bank, *Ethical Consumerism Report 2008*, available at: www.good withmoney.co.uk/ethical-consumerism-report-08/; Fairtrade Foundation, *Annual Report 2008–2009*, available at: www.fairtrade.org.uk/annual_review/awareness_ and_sales/.

3 'Voting with your trolley', *The Economist*, 9 December 2006, pp. 73–75.

4 'Good food?' *The Economist*, 9 December 2006, p. 12.

5 M. Lynas, 'Can shopping save the planet?' *The Guardian*, 17 September 2007, pp. 4–7.

6 Simon Billing, 'Fairtrade Logo amplified in UK with Cadbury's deal', see www.twin.org.uk/resources/news?n=22934 (accessed 15 September 2009).

CHAPTER TWO: THE ETHICAL PROBLEMATIZATION OF 'THE CONSUMER'

1 Richard Reeves (2003) *The Politics of Happiness*, NEF discussion paper. London: New Economics Foundation, p. 9.

2   Food Ethics Council (2005) *Getting Personal: Shifting responsibilities for dietary health*. Brighton: Food Ethics Council.
3   The EPI has since been replaced by annual Ethical Consumerism Reports, using the same methodology.
4   Collins, J., Thomas, G., Willis, R. and Wilsdon, J. (2003) *Carrots, Sticks and Sermons: Influencing public behaviour for environmental goals*. London: Demos/ Green Alliance, p. 49.
5   National Consumer Council (2007) *The Environmental Contract: How to harness public action on climate change*. London: NCC, p. 14.

CHAPTER THREE: PRACTISING CONSUMPTION

1   *Divine* was the first fair-trade chocolate bar to be available in the UK. It was an initiative of the Day Chocolate Company, jointly owned by farmers in West Africa and Twin Trading, the fair trade company set up by the Greater London Council in the 1980s.
2   For further information on this programme, see www.consume.bbk.ac.uk/
3   Levett, R., Christie, I., Jacobs, M. and Therivel, R. (2003) *A Better Choice of Choice: Quality of life, consumption and economic growth*, Fabian Policy Report 58. London: Fabian Society.

CHAPTER FOUR: PROBLEMATIZING CONSUMPTION

1   Cowe, R. and Williams, S. (2002) *Who Are the Ethical Consumers?* Co-operative Bank, p. 11.
2   Collins, J., Thomas, G., Willis, R. and Wilsdon. J. (2003) *Carrots, Sticks and Sermons: Influencing public behaviour for environmental goals*. London: Demos and Green Alliance.
3   Khaneka, P. (2004) *Do the Right Things! A practical guide to ethical living*. Oxford: New Internationalist Publications; Clark, D. (2004) *The Rough Guide to Ethical Shopping*. London: Rough Guides.
4   *The Ecologist* (2001) *Go Mad! 365 Daily Ways to Save the Planet*. London: Think Publishing.
5   Elkington, J. and Hailes, J. (1988) *The Green Consumer Guide: From shampoo to champagne – high street shopping for a better environment*. London: Guild, pp. 1–2.
6   Ethical Consumer Research Association (n.d.) *An Introduction to Ethical Consumer*. Manchester: ECRA Publishing Ltd, p. 5.
7   Focus Group, Bristol, 6 September 2005.
8   Ethical Marketing Group (2002) *The Good Shopping Guide*. London: Ethical Marketing Group, p. 11.
9   Ethical Consumer Research Association (n.d.) *An Introduction to Ethical Consumer*. Manchester: ECRA Publishing Ltd, p. 5.
10 Roddick, A. (2001) *Taking It Personally: How globalization affects you and powerful ways to challenge it*. London: Element, pp. 9–10.

11   Ibid.: 76.
12   Young, W. and Welford, R. (2002) *Ethical Shopping: Where to shop, what to buy and what to do to make a difference*. London: Fusion Press, pp. 4–5.
13   Ibid.: 44.
14   Ethical Consumer Research Association (n.d.) *An Introduction to Ethical Consumer*. Manchester: ECRA Publishing Ltd, p. 4.
15   Ibid.
16   Ibid.
17   Ethical Marketing Group (2002) *The Good Shopping Guide*. London: Ethical Marketing Group, p. 9.
18   Elkington, J. and Hailes, J. (1988) *The Green Consumer Guide: From shampoo to champagne – high street shopping for a better environment*. London: Guild, p. 4.
19   Ethical Consumer Research Association (n.d.) *An Introduction to Ethical Consumer*. Manchester: ECRA Publishing Ltd, p. 5.
20   National Consumer Council (2004) *Making Public Services Personal: A new compact for public services* (The Independent Policy Commission on Public Services report to the National Consumer Council). London: National Consumer Council, p. 10.
21   Ibid.
22   Boyle, D. and Simms, A. (2001) *The Naked Consumer: Why shoppers deserve honest product labelling*. London: New Economics Foundation, p. 11.
23   Gordon, W. (2002) *Brand Green: Mainstream or forever niche?* London: Green Alliance.
24   Williams, S., Doane, D. and Howard, M. (2003) *Ethical Consumerism Report 2003*. The Co-operative Bank, p. 7.
25   Ibid.: 4.
26   Doane, D. and Williams, S. (2002) *Ethical Purchasing Index 2002*. The Co-operative Bank, p. 5.
27   Ibid.
28   MacGillivray, A. (2000) *The Fair Share: The growing market share of green and ethical products*. London: New Economics Foundation, p. 2.
29   Hines, C. and Ames, A. (2000) *Ethical Consumerism: A research study*. MORI and the Co-operative Bank.
30   Corrado, M. and Hines, C. (2001) 'Business ethics: Making the world a better place', paper presented at the Market Research Society Conference, Brighton, p. 2.
31   National Consumer Council (2004) *Making Public Services Personal: A new compact for public services* (The Independent Policy Commission on Public Services report to the National Consumer Council). London: National Consumer Council, p. 43.
32   Ibid.: 61.
33   Richard Reeves (2003) 'The politics of happiness', NEF discussion paper. London: New Economics Foundation, pp. 4–5.
34   See also Holdsworth, M. (2003) *Green choice: What choice?* London: National Consumer Council.
35   Levett, R., with Christie, I., Jacobs, M. and Therivel, A. (2003) *A Better Choice of Choice: Quality of life, consumption and economic growth*. London: Fabian Society, p. 45.

36   Ibid.
37   Collins, J., Thomas, G., Willis, R. and Wilsdon. J. (2003) *Carrots, Sticks and Sermons: Influencing public behaviour for environmental goals.* London: Demos and Green Alliance.
38   Sustainable Consumption Roundtable (2006) *I will if you will: Towards sustainable consumption.* London: Sustainable Development Commission and the National Consumer Council.
39   *The Guardian,* 2 May 2006.
40   Williams, S., Doane, D. and Howard, M. (2003) *Ethical Consumerism Report 2003.* The Co-operative Bank, p. 6.
41   Hickman, L. (2005) *A Life Stripped Bare: Tiptoeing through the ethical minefield.* London: Transworld Publishers.

CHAPTER FIVE: GRAMMARS OF RESPONSIBILITY

1   Focus group, Bishopston, Bristol, 23 April 2004.
2   See *Oxford Commission on Sustainable Consumption Report,* 2004, Mansfield College, Oxford, pp. 37–44.
3   In addition to these ten focus groups, in 2005 we also convened two focus groups composed of 'committed' ethical consumers, who were recruited by placing an advert in *Ethical Consumer* magazine.
4   Ashley focus group, 6 April 2004.
5   Southmead focus group, 24 February 2004.
6   Knowle focus group, 26 February 2004.
7   Hartcliffe focus group, 8 March 2004.
8   Windmill Hill focus group, 27 November 2003.
9   Henleaze focus group, 6 May 2004.
10   Ashley focus group, 6 April 2004.
11   Easton focus group, 24 February 2004.
12   Windmill Hill focus group, 27 November 2003.
13   Stockwood focus group, 25 March 2004.
14   Bishopston focus group, 23 April 2004.
15   Knowle focus group, 26 February 2004.
16   Bishopston focus group, 23 April 2004.
17   Bishopston focus group, 23 April 2004.
18   Easton focus group, 24 February 2004.
19   Windmill Hill focus group, 27 November 2003.
20   Bishopston focus group, 23 April 2004.
21   Hartcliffe focus group, 8 March 2004.
22   Hartcliffe focus group, 8 March 2004.
23   Hartcliffe focus group, 8 March 2004.
24   Ashley focus group, 6 April 2004.
25   Stockwood focus group, 24 March 2004.
26   Stockwood focus group, 24 March 2004.
27   Henleaze focus group, 6 May 2004.
28   Henleaze focus group, 6 May 2004.

CHAPTER SIX: LOCAL NETWORKS OF GLOBAL FEELING

1   See www.european-fair-trade-association.org/Efta/Doc/What.pdf (accessed 18 August 2006).
2   See www.fairtrade.org.uk (accessed 7 December 2009).
3   *Traidcraft's Strategy 2002–2005*. Gateshead: Traidcraft, 2002.
4   See www.traidcraft.co.uk (accessed September 2005).
5   Interview with Mike Gidney, Director of Policy, Traidcraft, September 2005.
6   Interview with Stuart Palmer, Director of Marketing, Traidcraft, October 2004.
7   See www.traidcraftinteractive.co.uk.
8   Interview with Brian Conder, Key Accounts Manager, Traidcraft, October 2004.
9   Ibid.
10 Interview with Emma, Traidcraft Fairtrader, April 2005.
11 Interview with Brian Conder, Key Accounts Manager, Traidcraft, October 2004.
12 Interview with Sarah, Traidcraft customer, June 2005.
13 Interview with Liz, Traidcraft customer, May 2005.
14 Ibid.
15 Interview with Erika, Traidcraft customer and activist, July 2005.
16 Ibid.
17 Interview with Sue, Traidcraft Fairtrader, April 2005.
18 Interview with Edna, Traidcraft Fairtrader, April 2005.
19 Interview with Debbie, Traidcraft Key Contact, April 2004.
20 Interview with Sarah, Traidcraft customer, June 2005.
21 Interview with Hattie, Traidcraft Fairtrader, June 2005.
22 Ibid.
23 Ibid.
24 Interview with John, Traidcraft customer, June 2005.
25 Interview with Pauline, Traidcraft customer and activist, July 2005.
26 Interview with Brian Conder, Key Accounts Manager, Traidcraft, October 2004.
27 Interview with Stuart Palmer, Director of Marketing, Traidcraft, October 2004.
28 Interview with Sarah, Traidcraft customer, June 2005.
29 Interview with Liz, Traidcraft customer, May 2005.
30 Ibid.
31 Interview with Debbie, Traidcraft Key Contact, October 2004.
32 Interview with John, Traidcraft customer, June 2005.
33 Ibid.
34 Interview with Jennie, Traidcraft Fairtrader, June 2005.
35 Ibid.
36 Interview with Pauline, Traidcraft customer and activist, July 2005.
37 Interview with Edna, Traidcraft Fairtrader, April 2005.
38 Ibid.
39 Café Unlimited closed at the end of 2006, but this has not been the end of Jennie's involvement in fair trade campaigning. Previously active in the Fairtrade City Campaign (see Chapter 7) as a representative of local business, she went

on to become coordinator of the Bristol Fairtrade Network, responsible for extending fair trade principles once accreditation as a Fairtrade City was awarded in 2005.

40 Interview with Erika, Traidcraft customer and activist, July 2005.

CHAPTER SEVEN: FAIRTRADE URBANISM

1 *Fairtrade Foundation Annual Review 2008–2009*, available at: www.fairtrade.org.uk/annual_review/campaigns/?dm_i=605304445 (accessed 14 December 2009).

2 The ethnographic aspect of this empirical analysis is based primarily on the involvement of one of the authors (Alice Malpass) with the Fairtrade City campaign steering committee. This participant observation was supplemented by in-depth interviews with key local actors, including other steering group members, and participant observation by all of the authors in various local campaigning events and online discussion groups over the course of the year-long Fairtrade City campaign in Bristol. The ESRC/AHRC-funded *Governing the Subjects and Spaces of Ethical Consumption* research project also helped to finance the formal launch event of the campaign in June 2004, including hosting a two-week long Fairtrade Foundation photography exhibition at the CREATE Centre in Bristol. This event was used to undertake participant observation research, to network with local policy makers and activists, and as an access point for the year-long participant observation study led by Alice of the dynamics of the successful fair trade campaign.

3 Steven Morris, 'From slave trade to fair trade: Bristol's new image', *The Guardian*, 4 March 2005; James Russell, 'From slave trade to fair trade', *Bristol Fairtrade Directory 2007*, available at: www.bristolfairtradenetwork.org.uk.

4 Fairtrade Foundation (2002) *Fairtrade Town: Goals and Action Guide*. London: FTF.

5 Ibid.

6 Office of Government Commerce (2004) *Guidance on Fair and Ethical Trading*, p. 1, available at: www.ogc.gov.uk/social_issues_in_purchasing_fair_and_ethical_trade.asp (accessed 18 January 2005).

7 Interview with Corporate Procurement Manager, Bristol City Council, 2004.

8 Interview with Corporate Procurement Manager, Bristol City Council, 2004.

9 *Bristol Fairtrade City Challenge*, 2004. Bristol: Bristol Fairtrade Network.

10 Interview with local authority officer, Bristol Fairtrade City steering committee, 2004.

11 Sherrie Eugene, in *Bristol Fairtrade City Challenge*, 2004. Bristol: Bristol Fairtrade Network.

12 Interview with local authority officer, Bristol Fairtrade City steering committee, 2004.

13 Fairtrade Foundation (2002) *Fairtrade Town: Goals and Action Guide*. London: FTF.

14 Ibid., p. 9.

15 Fairtrade Town online forum, available at: http://groups.yahoo.com/group/FairtradeTown/ (accessed 3 May 2004).

16 Ibid.

17 Interview with local authority officer, coordinator of Bristol Fairtrade City steering group and former employee of the FTF, 2004.

18 Interview with local authority officer, City Sustainability Team, member of the Bristol Fairtrade City steering group, 2004.

19 Bruce Crowther, Fairtrade Town online forum, available at: http://groups. yahoo.com/group/FairtradeTown/ (accessed 3 May 2004).

20 Interview with local authority officer, Bristol Fairtrade City steering committee, 2004.

21 Mark Rowe, 'The last great British high street', *The Independent*, 8 August 2004.

22 Interview with local authority officer, Bristol Fairtrade City steering committee, 2004.

References

Adams, M. and Raisborough, J. (2008) 'What can sociology say about fair trade?' *Sociology* 42, 1165–1182.

Allen, A. (2009) Discourse, power and subjectivation: the Foucault/Habermas debate reconsidered, *The Philosophical Forum* 40 (1): 1–29.

Allen, J. (2008) 'Claiming connections: a distant world of sweatshops?' in Barnett, C., Robinson, J., and Rose, G. (eds), *Geographies of Globalization: Living in a demanding world*. London: Sage, pp. 7–54.

Amin, A., Cameron, A. and Hudson, R. (2002) *Placing the Social Economy*. London: Routledge.

Amin, A. and Thrift, N. (2005) 'What's left? Just the future', *Antipode* 37 (2): 220–38.

Anderson, E. (2000) 'Beyond homo economicus: new developments in theories of social norms', *Philosophy and Public Affairs* 29 (2): 170–200.

Anderson, M. (2009) NGOs and Fair Trade: the social movement behind the label, in N. Crowson, M. Hilton and J. McKay (eds), *NGOs in Contemporary Britain: Non-state actors in society and politics since 1945*. Basingstoke: Palgrave Macmillan.

Andrews, G. (2005) 'The slow food story', *Soundings* 31: 88–102.

Austin, J. L. (1961) *Philosophical Papers*. Oxford: Oxford University Press.

Bacon, C. (2005) 'Confronting the coffee crisis: can Fair Trade, organic, and speciality coffees reduce small-scale farmer vulnerability in northern Nicaragua?' *World Development* 33: 497–511.

Bang, H. and Sorensen, E. (1999) 'The everyday maker: A new challenge to democratic governance', *Administrative Theory & Praxis* 21 (3): 325–342.

Barber, B. (2007) *Consumed: How markets corrupt children, infantilize adults, and swallow citizens whole*. New York: W.W. Norton.

Barnett, C. (2008) 'Political affect in public space: normative blind-spots in non-representational ontologies', *Transactions of the Institute of British Geographers* NS, 33: 186–200.

Barnett, C. (2009) 'Publics and markets: what's wrong with neoliberalism?' in S. Smith, S. Marston, R. Pain, and J. P Jones III (eds), *The Handbook of Social Geography*. London: Sage.

Barnett, C. (2010) 'Fair trade', in B. Bevir (ed.), *The Encyclopedia of Political Theory*. London: Sage.

Barnett, C., Cafaro, P. and Newholm, T. (2005) 'Philosophy and Ethical Consumption', in R. Harrison, T. Newholm and D. Shaw (eds), *The Ethical Consumer*. London: Sage, pp. 11–24.

Barnett, C., Cloke, P., Clarke, N. and Malpass, A. (2005) 'Consuming ethics: articulating the subjects and spaces of ethical consumption', *Antipode* 37: 23–45.

Barnett, C., Robinson, J. and Rose, G. (eds) (2008) *Geographies of Globalisation: Living in a demanding world*. London: Sage.

Barr, S. (2003) 'Strategies for sustainability: citizens and responsible environmental behaviour', *Area* 35: 227–240.

Barr, S. and Gilg, A. (2006) 'Sustainable lifestyles: framing environmental action in and around the home', *Geoforum* 37: 906–920.

Barrientos S. and Smith S. (2007) 'Mainstreaming fair trade in global value chains: own brand sourcing of fruit and cocoa in U.K. supermarkets. In L. Raynolds, D. Murray and J. Wilkinson (eds), *Fair Trade: The challenges of transforming globalization*. Routledge.

Bateman, C., Fraedrich, J. and Iyer, R. (2002) 'Framing effects within the ethical decision making processes of consumers', *Journal of Business Ethics* 36: 119-140.

Bauman, Z. (1993) *Postmodern Ethics*. Oxford: Blackwell.

Bauman, Z. (1995) *Life in Fragments*. Oxford: Blackwell.

Bauman, Z. (1999) *In Search of Politics*. Cambridge: Polity Press.

Bauman, Z. (2000) *Liquid Modernity*. Oxford: Blackwell.

Bauman, Z. (2001) 'Consuming life', *Journal of Consumer Culture* 1 (1): 9–29.

Bauman, Z. (2007) *Consuming Life*. Cambridge: Polity Press.

Bauman, Z. (2008) 'Exit *Homo politicus*, enter *Homo consumens*', in K. Soper and F. Trentmann (eds), *Citizenship and Consumption*. New York: Palgrave Macmillan, pp. 139–153.

Becchetti, L. and Huybrechts, B. (2008) 'The dynamics of fair trade as a mixed-form market', *Journal of Business Ethics* 81 (4): 733–750.

Beck, U. (1996) *The Reinvention of Politics*. Oxford: Blackwell.

Beck, U. (1999) *What is globalization?* Oxford: Blackwell.

Beck, U. (2006) *Cosmopolitan Vision*. Cambridge: Polity Press.

Beck, U. and Beck-Gernsheim, E. (2001) *Individualization*. London: Sage.

Beck, U., Giddens, A. and Lash, S. (1994) *Reflexive Modernization*. Cambridge: Polity Press.

Belk, R. W. (1988) 'Possessions and the extended self', *Journal of Consumer Policy* 15: 139–168.

Bennett, J. (2001) 'Commodity fetishism and commodity enchantment', *Theory and Event* 5 (1).

Bennett, W. (2004) 'Branded political communication: lifestyle politics, logo campaigns and the rise of global citizenship', in M. Micheletti, A. Føllesdal and D. Stolle (eds), *Politics, Products and Markets: Exploring political consumerism, past and present*. New Jersey: Transaction Publishers, pp. 101–126.

Berry, H. and McEachern, M. (2005) 'Informing ethical consumers', in R. Harrison, T. Newholm and D. Shaw (eds), *The Ethical Consumer*. London: Sage, pp. 69–87.

Bevir, M. and Trentmann, F. (2007) 'Civic choices: retrieving perspectives on rationality, consumption and citizenship', in K. Soper and F. Trentmann (eds), *Citizenship and Consumption*. Basingstoke: Macmillan pp. 19–33.

Bickerstaff, K. and Walker, G. (2002) 'Risk, responsibility, and blame: an analysis of vocabularies of motive in air-pollution(ing) discourses', *Environment and Planning A* 34: 2175–2192.

Billig, M. (1991) *Ideology and Opinions: Studies in rhetorical psychology.* London: Sage.

Billig, M. (1996) *Arguing and Thinking.* Cambridge: Cambridge University Press.

Binkley, S. (2006) 'The perilous freedoms of consumption: toward a theory of the conduct of consumer conduct', *Journal of Cultural Research* 10, 343–362.

Bird, K. and Hughes, D. (1997) 'Ethical consumerism: the case of 'fairly traded' coffee, *Business Ethics* 6 (3): 159–167.

Bloch, M. (1998) *How We Think They Think.* Boulder, CO: Westview.

Blowfield M and Gallet S (2000) *Volta River Estates Fair Trade Bananas Case Study.* Ethical Trade and Sustainable Livelihoods Case Study Series. London: University of Greenwich.

Boltanski, L. and Thévenot, L. (2000) 'The reality of moral expectations: a sociology of situated judgement'. *Philosophical Explorations* 3: 208–231.

Boltanski, L. and Thévenot, L. (2006) *On Justification: economies of worth.* Princeton: Princeton University Press.

Bondi, L. (2005) 'Working the spaces of neoliberal subjectivity', *Antipode* 37: 497–514.

Bridge, G. (2005) *Reason in the City of Difference.* London: Routledge.

Bryant, R. and Goodman, M. (2004) 'Consuming narratives: the political ecology of "alternative" consumption', *Transactions of the Institute of British Geographers NS* 29: 344–366.

Bulkeley, H. and Gregson, N. (2009) 'Crossing the threshold: municipal waste policy and household waste generation', *Environment and Planning A* 41: 929–945.

Burgess, J. (2003) 'Sustainable consumption: is it really achievable?' *Consumer Policy Review* 13 (3): 78–84.

Burgess, J., Limb, M. and Harrison, C. M. (1988) 'Exploring environmental values through the medium of small groups: 1. Theory and practice', *Environment and Planning A* 20: 309–326.

Campbell, C. (1990) *The Romantic Ethics and the Spirit of Modern Consumerism.* Oxford: Basil Blackwell.

Carter, N. and Huby, M. (2005) 'Ecological citizenship and ethical investment', *Environmental Politics* 14: 255–272.

Cherrier, H. and Murray, J. (2002) 'Drifting away from excessive consumption: a new social movement based on identity construction', *Advances in Consumer Research* 29: 245–247.

Clarke, J., Newman, J., Smith, N., Vidler, E. and Westmareland, L. (2007) *Creating Citizen-Consumers.* London: Sage.

Clarke, M. (2010) *Challenging Choices: Ideology, consumerism and policy.* Bristol: Policy Press.

Clarke, N. (2008) 'From ethical consumerism to political consumption', *Geography Compass* 2 (6): 1870–1884.

Clarke, N., Cloke, P., Barnett, C. and Malpass, A. (2008) 'The spaces and ethics of organic food', *Rural Studies,* 24: 219–230.

Clavin, B. and Lewis, A. (2005) 'Focus groups on consumers' ethical beliefs', in R. Harrison, T. Newholm and D. Shaw (eds) *The Ethical Consumer.* London: Sage.

Cloke, P., Barnett, C., Clarke, N. and Malpass, A. (2010) 'Faith in ethical consumption', in L. Thomas (ed.), *Religion, Consumerism, and Sustainability: Paradise lost?* Basingstoke and New York: Palgrave Macmillan.

Cloke, P., Cook, I., Crang, P., Goodwin, M., Painter, J. and Philo, C. (2004) *Practising Human Geography.* London: Sage.

Clough, P. T. (2009) 'The new empiricism: affect and sociological method', *European Journal of Social Theory* 12: 43–61.

Cohen, G. A. (2000) *If You're An Egalitarian, How Come You Are So Rich?* Cambridge, MA: Harvard University Press.

Cohen, G. A. (2008) *Rescuing Justice and Equality.* Cambridge, MA: Harvard University Press.

Colls, R. and Evans, B. (2008) 'Embodying responsibility: children's health and supermarket initiatives', *Environment and Planning A*, 40: 615–631.

Connell, R. (2007) *Southern Theory.* Polity: Cambridge.

Cook, I., Harrison, M. and Lacey, C. (2006) 'The power of shopping', in R. Wilson (ed.), *Post Party Politics: can participation reconnect people and government?* London: Involve.

Coşgel, M. and Minkler, L. (2004) 'Religious identity and consumption', *Review of Social Economy* 62 (3): 339–350.

Cox, K. (1998) 'Spaces of dependence, spaces of engagement and the politics of scale: or, looking for local politics', *Political Geography* 17: 1–23.

Crane, A. (1999) 'Are you ethical? Please tick yes or no. On researching ethics in business organizations', *Journal of Business Ethics* 20: 237–248.

Crane, A. (2001) 'Unpacking the ethical product', *Journal of Business Ethics* 30: 361–373.

Crang, M. and Graham, S. (2007) 'Sentient cities: ambient intelligence and the politics of urban space', *Information, Communication & Society* 10: 789–817.

Crocker, D. and Linden, T. (eds) (1998) *Ethics of Consumption: the good life, justice and global stewardship.* London: Rowman & Littlefield.

Crossley, N. (2007) 'Social networks and extra-parliamentary politics', *Sociology Compass* 1 (1): 222–236.

Dauvergne, P. (2008) *The Shadows of Consumption: Consequences for the global environment.* Cambridge, MA: MIT Press.

Davies, B. and Harré, R. (1990) 'Positioning: the discursive production of selves', *Journal for the Theory of Social Behaviour* 20: 43–63.

Day Sclater, S. (2005) 'What is the subject?' *Narrative Inquiry* 13: 317–330.

De Pelsmacker, P., Driesen, L. and Rayp, G. (2005) 'Do consumers care about ethics? Willingness to pay for fair-trade coffee', *Journal of Consumer Affairs* 39: 363–385.

Dean, M. (1999) *Governmentality: Power and rule in modern society.* London: Sage.

Dean, M. (2007) *Governing Societies.* Buckingham: Open University Press.

Della Porta, D. and Tarrow, S. (eds) (2005) *Transnational Protest and Global Activism: People, passions, and power.* Lanham, MD: Rowman and Littlefield.

Desforges, L. (2004) 'The formation of global citizenship: international non-governmental organisations in Britain', *Political Geography* 23: 549–569.

Diani, M. (2005) 'Cities in the world: local civil society and global issues in Britain', in D. Della Porta and S. Tarrow (eds), *Transnational Protest and Global Activism: People, passions, and power.* Lanham, MD: Rowman and Littlefield, pp. 45–67.

Dickinson, R. and Carsky, M. (2005) 'The consumer as economic voter', in R. Harrison, T. Newholm and D. Shaw (eds), *The Ethical Consumer*. London: Sage, pp. 25–36.

Doherty, B. and Tranchell, S. (2005) 'New thinking in international trade? A case study of the Day Chocolate Company', *Sustainable Development* 13 (3): 166–176.

Dreyfus, H. (2007a) 'The return of the myth of the mental', *Inquiry* 50 (4): 352–365.

Dreyfus, H. (2007b) 'Response to McDowell', *Inquiry* 50 (4): 371–377.

Dreyfus, H. and Dreyfus, S. (2005) 'Expertise in real world contexts', *Organization Studies* 26: 779–792.

Eagleton, T. (2003) *After Theory*. London: Penguin.

Eden, S., Bear, C. and Walker, G. (2007) 'Mucky carrots and other proxies: problematising the knowledge-fix for sustainable and ethical consumption', *Geoforum* 39: 1044–1057.

Edwards, D. (2006) 'Discourse, cognition and social practices: the rich surface of language and social interaction', *Discourse Studies* 8 (1).

Edwards, D. and Stokoe, E. H. (2004) 'Discursive psychology, focus group interaction and participants' categories', *British Journal of Developmental Psychology* 22: 499–507.

Egels-Zandén, N. and Hyllman, P. (2006) 'Exploring the effects of union–NGO relationships on corporate responsibility: the case of the Swedish Clean Clothes Campaign', *Journal of Business Ethics* 64: 303–316.

Elliott, A. and Lemert, C. (2006) *The New Individualism: The emotional costs of globalization*. London: Routledge.

Elster, J. (1989) *The Cement of Society*. Cambridge: Cambridge University Press.

Escobar, A. (2001) 'Culture sits in places', *Political Geography*, 20: 139–174.

Ewald, F. (1991) 'Norms, discipline, and the law', in R. Post (ed.) *Law, Order, Culture*. Berkeley: University of California Press, pp. 138–161.

Feher, M. (2007) *Nongovernmental politics*. New York: Zone Books.

Fine, B. (2002) *The World of Consumption: The material and cultural revisited* (2nd edn). London: Routledge.

Flyvbjerg, B. (2001) *Making Social Science Matter*. Cambridge: Cambridge University Press.

Foster, R. J. (2005) 'Commodity futures: labour, love and value', *Anthropology Today* 21 (4): 8–12.

Foucault, M. (1985) *The Uses of Pleasure: The history of sexuality, Volume Two*. London: Penguin Books.

Foucault, M. (1997) *Ethics. The essential works of Foucault 1954–1984, Volume One*. London: Penguin Books.

Foucault, M. (2000) *Power: The essential works of Foucault 1954–1984, Volume Three*. London: Penguin Books.

Fraser, N. (1997) *Justice Interruptus*. London: Routledge.

Freidberg, S. (2003) 'Cleaning up down South: supermarkets, ethical trade and African horticulture', *Social and Cultural Geography* 4: 27–43.

Freidberg, S. (2004) 'The ethical complex of corporate food power', *Environment and Planning D: Society and Space* 22: 513–531.

Gabriel, Y. and Lang, T. (1995) *The Unmanageable Consumer*. London: Sage.

Gamson, W. and Wolfsfeld, G. (1993) 'Movements and media as interacting systems', *Annals of the American Association of Political and Social Sciences* 528 (1): 114–125.

Giddens, A. (1984) *The Constitution of Society*. Cambridge: Polity Press.

Giddens, A. (1991) *Modernity and Self-Identity*. Cambridge: Polity Press.

Giddens, A. (1994) *Beyond Left and Right: The future of radical politics*. Cambridge: Polity Press.

Giddens, A. (1998) *The Third Way: the renewal of social democracy*. Cambridge: Polity Press.

Ginsborg, P. (2005) *The Politics of Everyday Life: Making choice, changing lives*. New Haven, CT: Yale University Press.

Global Action Plan. (2004) *Consuming Passion – Do We Have To Shop Until We Drop?* London: Global Action Plan.

Gökariksel, B. and Mitchell, K. (2005) Veiling, secularism, and the neoliberal subject: national narratives and supranational desires in Turkey and France', *Global Networks* 5 (2): 147–165.

Golding, K. and Peattie, K. (2005) 'In search of a golden blend: perspectives on the marketing of fair trade coffee', *Sustainable Development* 13: 154–165.

Goodin, R. (2002) *Reflective Democracy*. Oxford: Oxford University Press.

Goodman, M. (2004) 'Reading fair trade: political ecological imaginary and the moral economy of fair trade foods', *Political Geography*, 23: 891–915.

Gould, N. (2003) 'Fair trade and the consumer interest: a personal account', *International Journal of Consumer Studies*, 27: 341–345.

Gregson, N. (2007) *Living With Things: Ridding, accommodation and dwelling*. Wantage: Sean Kingston Publishing.

Gregson, N. and Crewe, L. (2003) *Second Hand Cultures*. Oxford: Berg.

Gronow, J. and Warde, A. (2001) *Ordinary Consumption*. London: Routledge.

Grundy, J. and Smith, M. (2007) 'Activist knowledge in queer politics', *Economy and Society* 36, 294–317.

Guthman, J. (2007) 'The Polanyian way? Voluntary food labels as neoliberal governance', *Antipode* 39: 456–478.

Guthman, J. (2008) 'Thinking inside the neoliberal box: the micropolitics of agro-food philanthropy', *Geoforum* 39: 1241–1253.

Guthman, J. and DuPuis, M. (2006) 'Embodying neoliberalism: economy, culture and the politics of fat', *Environment and Planning D: Society and Space* 24: 3.

Habermas, J. (1989) *The Structural Transformation of the Public Sphere*. Cambridge: Polity Press.

Hacking, I. (1999) *The Social Construction of What?* Cambridge, MA: Harvard University Press.

Hacking, I. (2002) *Historical Ontology*. Cambridge, MA: Harvard University Press.

Hacking, I. (2004) 'Between Michel Foucault and Erving Goffman: between discourse in the abstract and face-to-face interaction', *Economy and Society* 33: 277–302.

Hajer, M. (1995) *The Politics of Environmental Discourse*. Oxford: Oxford University Press.

Hajer, M. and Versteeg, W. (2005) 'A decade of discourse analysis of environmental politics: achievements, challenges, perspectives', *Journal of Environmental Policy and Planning* 7 (3): 178–184.

Hale, A. and Wills, J. (eds) (2005) *Threads of Labour: garment industry supply chains from the workers' perspective*. Oxford: Blackwell.

Hamilton, T. (2009) 'Power in numbers: a call for analytical generosity toward new political strategies', *Environment and Planning A* 41: 284–301.

Hammersley, M. (1997) 'On the foundations of critical discourse analysis', *Language and Communication* 17 (3): 237–248.

Hammersley, M. (2003) 'Conversation analysis and discourse analysis: methods or paradigms?' *Discourse and Society* 14: 751–781.

Hanson, S. (2003) 'Geographical and feminist perspectives on entrepreneurship', *Geographische Zeitschrift* 91: 1–23.

Harré, R. (1991) 'The discursive production of selves', *Theory and Psychology* 1: 51–63.

Harré, R. (2002) 'Public sources of the personal mind: social constructionism in context', *Theory and Psychology* 12: 611–623.

Harré, R. (2004) 'Discursive psychology and the boundaries of sense', *Organization Studies* 25: 1435–1453.

Harré, R. and van Langenhove, L. (1991) 'Varieties of positioning', *Journal for the Theory of Social Behaviour* 21: 393–406.

Harrison, R. (2005) 'Pressure groups, campaigns and consumers', in R. Harrison, T. Newholm and D. Shaw (eds), *The Ethical Consumer*. London: Sage, pp. 55–67.

Harrison, R. (2008) 'The ethical consumption movement in the UK', paper presented at Workshop on Consumption Motivation, University of Reading Business School, 16 May.

Harrison, R., Newholm, T. and Shaw, D. (2005) *The Ethical Consumer*. London: Sage.

Hartwick, E. (2000) 'Towards a geographical politics of consumption', *Environment and Planning A* 32: 1177–1192.

Harvey, D. (2003) *A Short History of Neoliberalism*. Oxford: Oxford University Press.

Hermans, H. J. M. (2001) 'The construction of a personal position repertoire: method and practice', *Culture and Psychology* 7: 323–365.

Hetherington, K. (2004) 'Second-handedness: consumption, disposal and absent presence, *Environment and Planning D: Society and Space* 22 (1): 157–173.

Hetherington, K. (2007) *Capitalism's Eye: Cultural Spaces of the Commodity*. New York: Routledge.

Hilson, G. (2008) '"Fair trade gold": Antecedents, prospects and challenges', *Geoforum* 39 (1): 386–400.

Hilton, M. (2003) *Consumerism in Twentieth-Century Britain: The search for a historical movement*. Cambridge: Cambridge University Press.

Hilton, M. (2004) 'The legacy of luxury: moralities of consumption since the 18th century', *Journal of Consumer Culture* 4: 101–123.

Hilton, M. (2005) 'The duties of citizens, the rights of consumers', *Consumer Policy Review* 15 (1): 6–12.

Hilton, M. (2007a) 'Social activism in an age of consumption: the organized consumer movement', *Social History* 32 (2): 121–143.

Hilton, M. (2007b) 'The banality of consumption', in K. Soper and F. Trentmann (eds), *Citizenship and Consumption*. New York: Palgrave Macmillan.

Hilton, M. (2008) *Prosperity for All: Consumer activism in an era of globalization*. Ithaca: Cornell University Press.

Hinchliffe, S. (1996) 'Helping the earth begins at home: the social construction of environmental responsibilities', *Global Environmental Change* 6: 53–62.

Hinchliffe, S. (1997) 'Locating risk: energy use, the "ideal" home and the non-ideal world', *Transactions of the Institute of British Geographers* 22, 197–209.

Hinrichs, C. (2003) 'The practice and politics of food system localisation', *Journal of Rural Studies* 19: 23–32.

Hira, A. and Ferrie, J. (2006) 'Three challenges for reaching the mainstream', *Journal of Business Ethics* 63 (2): 107–118.

Hobson, K. (2002) 'Competing discourses of sustainable consumption: does the rationalisation of lifestyles make sense?' *Environmental Politics* 11 (2): 95–120.

Hobson, K. (2003) 'Thinking habits into action: the role of knowledge and process in questioning household consumption practices', *Local Environment* 8 (1): 95–112.

Hobson, K. (2006a) 'Bins, bulbs and shower timers: on the techno-ethics of sustainable living', *Ethics, Place and Environment* 9 (3): 335–354.

Hobson, K. (2006b) 'Environmental psychology and the geographies of ethical and sustainable consumption: aligning, triangulating, challenging?' *Area* 38 (3): 292–300.

Hodges, I. (2002) 'Moving beyond words: therapeutic discourse and ethical problematization', *Discourse Studies* 4: 455–479.

Holloway, L. and Kneafsey, M. (2000) 'Reading the spaces of the farmers' market', *Sociologia Ruralis* 40: 285–299.

Honneth, A. (1991) *The Critique of Power*. Cambridge, MA: MIT Press.

Hooghe, M. and Micheletti, M. (2005) 'Politics in the supermarket: political consumerism as a form of political participation', *International Political Science Review* 26 (3): 245–269.

Hudson, I. and M. Hudson. (2003) 'Removing the veil: commodity fetishism, fair trade, and the environment', *Organization and Environment* 16 (4): 413–430.

Hughes, A. (2001) 'Global commodity networks, ethical trade and governmentality: organising business responsibility in the Kenyan cut flower industry', *Transactions of the Institute of British Geographers* 26: 390–406.

Hughes, A. (2005) 'Geographies of exchange and circulation: alternative trading spaces', *Progress in Human Geography* 29: 496–504.

Hughes, A., Buttle, M. and Wrigley, N. (2007) 'Organisational geographies of corporate social responsibility: a UK–US comparison of retailers' ethical trading initiatives', *Journal of Economic Geography* 7: 491–513.

Hurley, S. L. (2003) *Justice, Luck and Knowledge*. Cambridge, MA: Harvard University Press.

Isin, E. (2002) *Being Political: Genealogies of citizenship*. Minneapolis: University of Minnesota Press.

Isin, E. and Nielsen, G. (eds) (2008) *Acts of Citizenship*. London: Macmillan.

Jackson, P. (2004) 'Local consumption cultures in a globalising world', *Transaction of the Institute of British Geographers*, 29: 165–185.

Jackson, P., Brooks, K. and Stevenson, N. (1999) 'Making sense of men's lifestyle magazines', *Environment and Planning D: Society and Space* 17: 353–368.

Jackson, P., Ward, N. and Russell, P. (2009) 'Moral economies of food and geographies of responsibility', *Transactions of the Institute of British Geographers NS* 34: 12–24.

Jackson, T. (2003) 'Sustainable consumption and symbolic goods', paper presented at ESRC Workshop, *Coming to Terms with Consumption*, Birkbeck College, October.

Jackson, T. (2004) *Motivating Sustainable Consumption: A review of evidence on consumer behaviour and behaviour change*. A report to the Sustainable Development Research Network, Policy Studies Institute, London.

Jacobsen, E. and Dulsrud, A. (2007) 'Will consumers save the world? The framing of political consumerism', *Journal of Agricultural and Environmental Ethics* 20: 469–482.

Jacques, M. (2004) 'The death of intimacy: a selfish, market-driven society is erod-ing our very humanity', *The Guardian*, September 18.

Johns, R. and Vural, L. (2000) 'Class, geography, and the consumerist turn: UNITE and the Stop Sweatshops campaign', *Environment and Planning A* 32 (7): 1193–1213.

King, M. (2009) 'Clarifying the Foucault/Habermas debate: morality, ethics, and "normative foundations"', *Philosophy and Social Criticism* 35 (3): 287–314.

Kitzinger, J. (1994) 'The methodology of focus groups: the importance of interac-tion between research participants', *Sociology of Health and Illness* 16: 103–121.

Klein, N. (2000) *No Logo*. London: Flamingo.

Kleine, D. (2005) 'Fairtrade.com versus Fairtrade.org: how Fairtrade organisations use the Internet', *Interacting with Computers* 17: 57–83.

Kocken, D. (2002) *The Impact of Fair Trade*. Maastricht: European Fair Trade Association.

Kuper, A. (ed.) (2005) *Global Responsibilities: who must deliver on human rights?* London: Routledge.

Kurtz, H. (2003) 'Scale frames and counter-scale frames: constructing the problem of environmental injustice', *Political Geography* 22: 887–916.

Lamb, H. (2008) *Fighting the Banana Wars and other Fairtrade Battles*. London: Rider Books.

Lang, T. and Gabriel, Y. (2005) 'A brief history of consumer activism', in Harrison, R., Newholm, T. and Shaw, D. (eds), *The Ethical Consumer*. London: Sage.

Larner, W. (1997) 'The legacy of the social: Market governance and the consumer', *Economy and Society* 26 (3): 373–399.

Larner, W. (2003) 'Neoliberalism?' *Environment and Planning D: Society and Space* 21: 509–512.

Larner, W., Le Heron, R. and Lewis, N. (2007) 'Co-constituting "After Neoliberalism": globalizing governmentalities and political projects in Aotearoa, New Zealand', in K. England and K. Ward (eds), *Neoliberalization: States, networks, people*. Oxford: Blackwell.

Lasch, C. (1979) *The Culture of Narcissism*. New York: W.W. Norton.

Laurier, E. (2010) 'How to feel things with words', in B. Anderson and P. Harrison (eds), *Taking Place: Non-representational theories and geography*. London: Berg.

Lave, J. (1988) *Cognition in Practice: Mind, mathematics and culture in everyday life*. Cambridge: Cambridge University Press.

Lawson, N. (2009) *All Consuming: How shopping got us into this mess and how we can find our way out*. London: Penguin.

Leclair, M. (2002) 'Fighting the tide: alternative trade organisations in the area of global free trade', *World Development* 30: 949–958.

Lemke, T. (2001) 'The birth of bio-politics: Michel Foucault's lecture at the Collège de France on neo-liberal governmentality', *Economy and Society* 30: 190-207.

Lemke, T. (2002) 'Foucault, governmentality and critique', *Rethinking Marxism* 14 (3): 49–64.

Levi, M. and Linton, A. (2003) 'Fair trade: a cup at a time?' *Politics and Society* 31 (3): 407–432.

Leyshon, A., Lee, R. and Williams, C. (2003) *Alternative Economic Spaces*. London: Sage.

Li, Y. and Marsh, D. (2008) 'New forms of political participation: searching for expert citizens and everyday makers', *British Journal of Political Science* 38 (2): 247–272.

Linehan, C. and J. McCarthy. (2001) 'Positioning in practice: understanding participation in the social world', *Journal of the Theory of Social Behaviour* 30: 435–453.

Linton, A. (2008) 'Ethical trade initiatives', *Globalizations* 5 (2): 227–229.

Linton, A., Liou, C. C. and Shaw, K. A. (2004) 'A taste of trade justice: marketing global social responsibility via fair trade coffee', *Globalizations* 1 (2): 223–246.

Littler, J. (2005) 'Beyond the boycott: anti-consumerism, cultural change and the limits of reflexivity', *Cultural Studies* 19 (2): 227–252.

Littler, J. (2008) *Radical Consumption: Shopping for change in contemporary culture.* Buckingham: Open University Press.

Lockie, S. (2002) '"The invisible mouth": mobilising "the consumer" in food-consumption networks', *Sociologia Ruralis*, 42 (4): 278–294.

Lockie, S. and Goodman, M. (2006) 'Neoliberalism, standardisation and the problem of space: competing rationalities of governance in fair trade and mainstream agri-environmental networks', in T. Marsden and J. Murdoch (eds), *Between the Local and the Global: Confronting complexity in the contemporary agri-food sector.* Oxford: Elsevier, pp. 95–120.

Low, W. and Davenport, E. (2005) 'Postcards from the edge: maintaining the 'alternative' character of fair trade', *Sustainable Development* 13: 143–153.

Low, W. and Davenport, E. (2007) 'To boldly go . . . exploring ethical spaces to re-politicise ethical consumption and fair trade', *Journal of Consumer Behaviour* 6: 336–348.

Luetchford, P. and De Neve, G. (eds) (2008) *Hidden Hands in the Market: ethnographies of fair trade, ethical consumption and corporate social responsibility.* Bingley: Emerald Group Publishing.

Macnaghten, P. (2003) 'Embodying the environment in everyday life practices', *Sociological Review* 51 (1): 63–84.

Macnaghten, P. and Jacobs, M. (1997) 'Public identification with sustainable development: investigating cultural barriers to participation', *Global Environmental Change* 7 (1): 5024.

Malins A and Blowfield M. (2000) *Fruits of the Nile: Fair trade processing case study.* Ethical Trade and Sustainable Livelihoods Case Study Series. London: University of Greenwich.

Marquand, D. (2004) *The Decline of the Public: The hollowing out of Citizenship.* Cambridge: Polity Press.

Marres, N. (2007) 'Dilemmas of home improvement: can clean energy technology mediate civic involvement in climate change?' in M. Feher, with G. Krikorian and Y. McKee (eds), *Nongovernmental Politics.* New York: Zone Books, pp. 368–385.

Marres, N. (2009) 'Testing powers of engagement: green living experiments, the ontological turn and the undoability of involvement', *European Journal of Social Theory* 12: 117–133.

Marsden, T. (2000) 'Food matters and the matter of food: towards a new governance', *Sociologia Ruralis* 41: 20–29.

Massey, D. (2006) 'Space, time and political responsibility in the midst of global inequality', *Erkunde* 60: 89–95.

Massey, D. (2007) *World City*. Cambridge: Polity Press.

Mattingly, C. (1990) 'Narrative reflections on practical actions: two learning experiments in reflective storytelling', in D. A. Schon (ed.), *The Reflective Turn*. New York: Teachers College Press.

McCarthy, T. (1993) *Ideals and Illusions: On reconstruction and deconstruction in contemporary critical theory*. Cambridge MA: MIT Press.

McDonald, R, Mead, N., Cheragi-Sohi, S., Bower, P., Whalley, D. and Roland, M. (2007) 'Governing the ethical consumer: identity, choice and the primary care medical encounter', *Sociology of Health and Illness* 29: 430–456.

McDowell, J. (2007a) 'What myth?' *Inquiry* 50 (4): 338–351.

McDowell, J. (2007b) 'Response to Dreyfus', *Inquiry* 50 (4): 366–370.

McNay, L. (2000) *Gender and Agency*. Cambridge: Polity Press.

Micheletti, M. (2003) *Political Virtue and Shopping: individuals, consumerism and collective action*. London: Palgrave.

Micheletti, M. and Follesdal, A. (2007) 'Shopping for human rights', *Journal of Consumer Policy* 30: 167–175.

Micheletti, M. and Stolle, D. (2007) 'Mobilizing consumers to take responsibility for global social justice', *Annals of the American Academy of Political and Social Science* 611: 157–175.

Micheletti, M., Follesdal, A. and Stolle, D. (eds) (2004) *Politics, Products and Markets: Exploring political consumerism past and present*. London: Transaction.

Miller, B. (2000) *Geography and Social Movements*. Minneapolis: University of Minnesota Press.

Miller, D. (1995) 'Consumption as the vanguard of history', in D. Miller (ed.) *Acknowledging Consumption*. London: Routledge.

Miller, D. (1998) *A Theory of Shopping*. Cambridge, Polity Press.

Miller, D. (2001a) *Consumption: Critical concepts in the social sciences* (Volume 1). London: Routledge.

Miller, D. (2001b) *Consumption: Critical concepts in the social sciences* (Volume 2). London: Routledge.

Miller, D. (2001c) 'The poverty of morality', *Journal of Consumer Culture* 1: 225–243.

Miller, D. (2009) *Stuff*. Cambridge: Polity Press.

Miller, P. (2001) 'Governing by numbers: why calculative practices matter', *Social Research* 68 (2): 379–396.

Miller, P. and O'Leary, T. (1987) 'Accounting and the construction of the governable person', *Accounting, Organizations and Society* 12: 235–265.

Miller, P. and Rose, N. (1997) 'Mobilising the consumer: assembling the subject of consumption', *Theory, Culture and Society* 14 (1): 1–36.

Moberg, M. (2005) 'Fair trade and Eastern Caribbean banana farmers: rhetoric and reality in the anti-globalization movement', *Human Organization* 64 (1): 4–15.

Moore, G. (2004) 'The fair trade movement: parameters, issues and future research', *Journal of Business Ethics* 53: 73–86.

Moore, G. and Beadle, R. (2006) 'In search of organisational virtue in business: agents, goods, practices, institutions, and environments', *Organisation Studies* 27: 369–389.

Morgan, K. and Morley, A. (2004) 'Creating sustainable food chains: tapping the potential of positive public procurement', in M. Thomas and M. Rhisart (eds), *Sustainable Regions*. Cardiff: Aureus Publishing.

Morgan, K. and Sonnino, R. (2007) 'The creative procurement of school meals in Italy and the UK', *International Journal of Consumer Studies* 31: 19–25.

Murray, R. (2004) 'The new political economy of public life', *Soundings* 27: 19–32.

Murray, D. and Raynolds, L. (2000) ''Alternative Trade in Bananas': Obstacles and opportunities for progressive social change in the global economy', *Agriculture and Human Values* 17: 65–74.

Myers, G. (1998) 'Displaying opinions: topics and disagreement in focus groups', *Language in Society* 27: 85–111.

Myers, G. and Macnaghten, P. (1998) 'Rhetorics of environmental sustainability: commonplaces and places', *Environment and Planning A* 30: 333–353.

Needham, C. (2003) *Citizen-Consumers: New Labour's marketplace democracy.* London: Catalyst.

Nelson, V. and Galvez, M. (2000) *Social Impact of Conventional Cocoa Trading on Forest-dependent People in Ecuador.* National Resources Institute, University of Greenwich.

Newholm, T. and Shaw, D. (2007) 'Studying the ethical consumer: a review of research', *Journal of Consumer Behaviour* 6: 253–270.

Newman, J. (2005) 'Enter the transformational leader: network governance and the micro-politics of modernization', *Sociology* 39: 717–734.

Newman, J. (2006) 'A politics of the public', *Soundings* 32: 162–176.

Newton, T. (1998) 'Theorizing subjectivity in organizations: the failure of Foucauldian studies?' *Organization Studies* 19 (3): 415–447.

Nicholls, A. (2002) 'Strategic options in fair trade retailing', *International Journal of Retail and Distribution Management* 30 (1): 6–17.

Nicholls, A. and Opal, C. (2005) *Fair Trade: Market-driven ethical consumption.* London: Sage.

Norris, P. (1998) *Democratic Phoenix.* Cambridge: Cambridge University Press.

Norris, P. (2007) 'Political participation', in C. Boix and S. Stokes (eds), *The Oxford Handbook of Comparative Politics.* Oxford: Oxford University Press.

O'Neill, O. (1996) *Towards Justice and Virtue.* Cambridge: Cambridge University Press.

O'Neill, O. (2000) *The Bounds of Justice.* Oxford: Oxford University Press.

Pahl, J. (1999) *Invisible Money: Family finances in the electronic economy.* Bristol: Policy Press.

Pattie, C., Seyd, P. and Whiteley, P. (2003a) 'Citizenship and civic engagement: attitudes and behaviour in Britain', *Political Affairs*, 51: 443–468.

Pattie, C., Seyd, P. and Whiteley, P. (2003b) 'Civic attitudes and engagement in modern Britain', *Parliamentary Affairs* 56: 616–633.

Pogge, T. (1994) 'An egalitarian law of the peoples', *Philosophy and Public Affairs* 23: 195–223.

Pogge, T. (2000) 'On the site of distributive justice: reflections on Cohen and Murphy', *Philosophy and Public Affairs* 29 (2): 137–169.

Pogge, T. (2001) 'Priorities of global justice', *Metaphilosophy* 32 (1/2): 6–24.

Potter, J. (2003) 'Discursive psychology: between method and paradigm', *Discourse and Society* 14: 783–794.

Potter, J. and Wetherell, M. (1990) 'Discourse: noun, verb or social practice?' *Philosophical Psychology* 3 (2/3): 205–18.

Potter, J., Edwards, D. and Wetherell, M. (2002) 'A model of discourse in action', *American Behavioral Scientist* 36: 383–401.

Prasad, M., Kimeldorf, H., Meyer, R. and Robinson, I. (2004) 'Consumers of the world unite: a market-based response to sweatshops', *Labour Studies Journal* 29 (3): 57–80.

Princen, T., Maniates, M. and Conca, K. (2003) 'Confronting consumption', in T. Princen, M. Maniates and K. Conca (eds), *Confronting Consumption*. Cambridge, MA: MIT Press, pp. 1–20.

Purdue, D., Diani, M. and Lindsay, I. (2004) 'Civic networks in Bristol and Glasgow', *Community Development Journal* 39: 277–288.

Putnam, R. (2000) *Bowling Alone: The collapse and revival of American community*. New York: Simon and Schuster.

Pykett, J. (2009) 'Personalisation and de-schooling: uncommon trajectories in contemporary education policy', *Critical Social Policy* 100 (29): 3.

Pykett, J., Cloke, P., Barnett, C., Clarke, N. and Malpass, A. (2010) 'Educating the global citizen', *Environment and Planning D: Society and Space* 28: 487–508.

Randall, D. (2005) An exploration of opportunities for the growth of the fair trade market, *Journal of Business Ethics* 56: 55–67.

Ransom D. (2005) 'Fair trade for sale', *New Internationalist* 377 (April), available at: http://live.newint.org/issue377/essay.htm (accessed 23 October 2006).

Rawls, J. (1972) *A Theory of Justice*. Oxford: Oxford University Press.

Rawls, J. (1999) *The Law of Peoples*. Cambridge, MA: Harvard University Press.

Raynolds, L. (2000) Re-embedding global agriculture: the international organic and fair trade movements', *Agriculture and Human Values* 17: 297–309.

Raynolds, L. (2002) 'Consumer/producer links in fair trade coffee networks', *Sociologia Ruralis* 42: 404–424.

Raynolds, L., Murray, D. and Wilkinson, J. (2007) *Fair Trade: The challenges of transforming globalization*. London: Routledge.

Reckwitz, A. (2002) 'Toward a theory of social practices: a development in culturalist theorizing', *European Journal of Social Theory* 5: 243–263.

Redfern, A and Snedker, P. (2002) *Creating Market Opportunities for Small Enterprises: Experiences of the fair trade movement*. Geneva: International Labour Office.

Renard, M. (2003) 'Fair trade: quality, market and conventions', *Journal of Rural Studies* 19 (1): 87–96.

Renard, M. (2005) 'Quality certification, regulation and power in fair trade', *Journal of Rural Studies* 21: 419–431.

Rice, R. (2001) 'Noble goals and challenging terrain: organic and fairtrade coffee movements in the global marketplace', *Journal of Agricultural and Environmental Ethics* 14: 39–66.

Ronchi, L. (2002) *The Impact of Fair Trade on Producers and Their Organisations: A case study with Coocafe in Costa Rica*. Brighton: University of Sussex.

Ronchi, L. (2003) *Monitoring Impact of Fair Trade Initiatives: A case study of Kuapa Kokoo and the Day Chocolate Factory*. London: Twin Trading.

Root, A. (2007) *Market Citizenship: Experiments in democracy and globalization*. London: Sage.

Rose, N. (1999) *Powers of Freedom*. Cambridge: Cambridge University Press.

Rose, N., O'Malley, P. and Valverde, M. (2006) 'Governmentality', *Annual Review of Law and Social Science* 2: 83–104.

Sassatelli, R. (2006) 'Virtue, responsibility and consumer choice: framing critical consumerism', in J. Brewer and F. Trentmann (eds), *Consuming Cultures, Global Perspectives*. Oxford: Berg, pp. 219–250.

Sassatelli, R. (2007) *Consumer Culture: History, theory and politics*. London: Sage.

Sayer, A. (2003) '(De)commodification, consumer culture, and moral economy', *Environment and Planning D: Society and Space* 21: 341–357.

Sayer, A. (2004) 'Restoring the moral dimension: acknowledging lay normativity' (online paper available at: http://www.lancs.ac.uk/fss/sociology/papers/sayer-restoring-moral-dimension.pdf).

Sayer, A. (2005) *The Moral Significance of Class*. Cambridge: Cambridge University Press.

Schatzki, T. R. (1996) *Social Practices: A Wittgensteinian approach to human activity and the social*. Cambridge: Cambridge University Press.

Schatzki, T. R. (2002) *The Site of the Social: A philosophical account of the constitution of social life and change*. University Park: Pennsylvania State University Press.

Schatzki, T., Knorr-Cetina, K. and von Savigny, E. (2001) The Practice Turn in Contemporary Theory. London: Routledge.

Schor, J. B. (2007) 'In defense of consumer culture: revisiting the consumption debates of the twentieth century', *American Academy of Political and Social Science* 611: 16–30.

Schudson, M. (2006) 'The troubling equivalence of citizen and consumer', *Annals of the American Academy of Political and Social Science* 608: 193–204.

Schudson, M. (2007) 'Citizens, consumers, and the good society', *Annals of the American Academy of Political and Social Science* 611: 236–249.

Seyfang, G. (2004) 'Consuming values and contested cultures: a critical analysis of the UK strategy for sustainable consumption and production', *Review of Social Economy* 62: 323–338.

Seyfang, G. (2005) 'Shopping for sustainability: can sustainable consumption promote ecological citizenship', *Environmental Politics* 14 (2): 290–306.

Shamir, R. (2008) 'The age of responsibilization: on market-embedded morality', *Economy and Society* 37: 1–19.

Shaw, D. and Newholm, T. (2002) 'Voluntary simplicity and the ethics of consumption', *Psychology and Marketing* 19: 167–185.

Shaw, D. and Shiu, E. (2003) 'Ethics in consumer choice: a multivariate modelling approach', *European Journal of Marketing*, 37: 10.

Shotter, J. (1985) 'Accounting for place and space', *Environment and Planning D: Society and Space* 3: 447–460.

Shotter, J. (1993a) *The Cultural Politics of Everyday Life*. Oxford: Oxford University Press.

Shotter, J. (1993b) *Conversational Realities*. London: Sage.

Shove, E. (2003) *Comfort, Cleanliness and Convenience: The social organisation of normality*. Oxford: Berg.

Shreck, A. (2005) 'Resistance, redistribution and power in the fair trade banana initiative', *Agriculture and Human Values* 22: 17–29.

Slater, D. (1997) *Consumer Culture and Modernity*. Cambridge: Polity Press.

Slocum, R. (2004) 'Polar bears and energy-efficient lightbulbs: strategies to bring climate change home', *Environment and Planning D: Society and Space* 22: 413–438.

Smart, C. and Neale, B. (1997) 'Good enough morality? Divorce and postmodernity', *Critical Social Policy* 17: 3–27.

Smith, D. M. (2000) *Moral Geographies: Ethics in a world of difference*. Edinburgh: Edinburgh University Press.

Smith, S. and Barrientos, S. (2005) 'Fair trade and ethical trade: are there moves towards convergence?' *Sustainable Development* 13: 190–198.

Sonnino, R. (2009) 'Quality food, public procurement and sustainable development: the school meal revolution in Rome', *Environment and Planning A* 41: 425–440.

Soper, K. (2004) 'Rethinking the "good life": the consumer as citizen', *Capitalism, Nature, Socialism* 15 (3): 111–116.

Soper, K. (2006) 'Conceptualising needs in the context of consumer politics', *Journal of Consumer Policy* 29: 355–372.

Stark, D. (2009) *The Sense of Dissonance*. Princeton: Princeton University Press.

Stark, D., Vedres, B. and Bruszt, L. (2006) 'Rooted transnational publics', *Theory and Society* 35: 323–349.

Stoker, G. (2006) *Why Politics Matters: Making democracy work*. London: Palgrave Macmillan.

Stolle, D., Hooghe, M. and Micheletti, M. (2005) 'Politics in the supermarket: political consumerism as a form of political participation', *International Political Science Review* 26: 245–269.

Strong, C. (1997) 'The problem of translating fair trade principles into consumer purchase behaviour', *Marketing Intelligence and Planning* 15 (1): 32–37.

Strydom, P. (1999) 'The challenge of responsibility for sociology', *Current Sociology* 47 (3): 65–82.

Tallantire, A. (2000) 'Partnerships in fair trade: reflections from a case study of Cafédirect', *Development in Practice* 10 (2): 166–177.

Tasioulas, J. (2005) 'Global justice without end?' in C. Barry and T. Pogge (eds), *Global Institutions and Responsibility: Achieving global justice*. Oxford: Blackwell, pp. 3–29.

Taylor, C. (1993) 'Engaged agency and background in Heidegger', in Charles B. Guignon (ed.), *The Cambridge Companion to Heidegger*. New York: Cambridge University Press, pp. 317–336.

Taylor, C. (1995) *Philosophical Arguments*. Cambridge, MA: Harvard University Press.

Taylor, C. (2000) 'What's wrong with foundationalism?' in M. Wrathall and J. Malpas (eds), *Heidegger, Coping, and Cognitive Science*. Cambridge, MA: MIT Press, pp. 115–134.

Taylor, P. (2005) 'In the market but not of it: fair trade coffee and Forest Stewardship Council certification as market-based social change', *World Development* 33: 129–147.

Taylor, P., Murray, D. and Raynolds, L. (2005) 'Keeping trade fair: governance challenges in the fair trade coffee initiative', *Sustainable Development* 13: 199–208.

Taylor, S. (2005) 'Self-narration as rehearsal: a discursive approach to the narrative formation of identity', *Narrative Inquiry* 15: 45–50.

Taylor, S. and Littlejohn, K. (2006) 'Biographies in talk: a narrative-discursive research proposal', *Qualitative Sociology Review* 11 (1): 22–38.

Thaler, R. H. and Sunstein, C. R. (2007) *Nudge: Improving decisions about health, wealth and happiness*. New Haven, CT: Yale University Press.

Thompson, C. J. and G. Coskuner-Balli (2007) 'Enchanting ethical consumerism', *Journal of Consumer Culture* 7: 275–303.

Thrift, N. (1996) *Spatial Formations*. London: Sage.

Thrift, N. (2007) *Non-representational Theory*. London: Routledge.

Thrift, N. (2008) 'The material practices of glamour', *Journal of Cultural Economy* 1: 9–23.

Tiffin, P. (2002) 'A chocolate-coated case for alternative international business', *Development in Practice* 12 (3/4): 383–397.

Tilly, C. (1994) 'Social movements as historically specific clusters of political performances', *Berkeley Journal of Sociology* 38: 1–30.

Tilly, C. (2006) *Why?* Princeton: Princeton University Press.

Tilly, C. (2008) *Credit and Blame*. Princeton: Princeton University Press.

Tormey, S. (2007) 'Consumption, resistance and everyday life: ruptures and continuities', *Journal of Consumer Policy* 30: 263–280.

Törrönen, J. (2001) 'The concept of subject position in empirical social research', *Journal for the Theory of Social Behaviour* 31: 313–329.

Transfair USA (2004) *Annual Report*. Oakland, CA: Transfair USA.

Traub-Werner, M. and Cravey, A. J. (2002) 'Spatiality, sweatshops and solidarity in Guatemala', *Social and Cultural Geography* 3: 383–401.

Trentmann, F. (ed.) (2005) *The Making of the Consumer: Knowledge, power and identity in the modern world*. Oxford: Berg.

Trentmann, F. (2006a) 'Consumption', in J. Merriman and J. Winter (eds), *Europe since 1914: Encyclopedia of the Age of War and Reconstruction*, Volume 2. Detroit: Charles Scribner, pp. 704–715.

Trentmann, F. (2006b) 'The modern genealogy of the consumer: meanings, identities and political synapses before affluence', in J. Brewer and F. Trentmann (eds), *Consuming Cultures, Global Perspectives*. Oxford: Berg, pp. 19–69.

Trentmann, F. (2008) *Free Trade Nation: Commerce, consumption and civil society in modern Britain*. Oxford: Oxford University Press.

Trentmann, F. and Morgan, B. (2006) 'Introduction: the politics of necessity', *Journal of Consumer Policy* 29: 345–353.

Tully, J. (1989) Wittgenstein and political philosophy: understanding practices of critical reflection, *Political Theory* 17: 172–204.

Van den Burg, S., Mol, A. and Spaargaren, G. (2003) 'Consumer-oriented monitoring and environmental reform', *Environment and Planning A* 21: 371–388.

Van Vliet, B., Chappells, H. and Shove, E. (2005) *Infrastructures of Consumption*. Oxford: Blackwell.

Varul, M. Z. (2008) 'Consuming the Campesino: fair trade marketing between recognition and romantic commodification', *Cultural Studies* 22 (5): 654–679.

Varul, M. Z. (2009) 'Ethical selving in cultural contexts: fairtrade consumption as an everyday ethical practice in the UK and in Germany', *International Journal of Consumer Studies*, 33 (2): 183–189.

Walkerdine, V. (2005) 'Freedom psychology and the neoliberal subject', *Soundings* 25: 47–61.

Warde, A. (2005) 'Consumption and theories of practice', *Journal of Consumer Culture* 5: 131–153.

Watson, M. (2007) 'Trade justice and individual consumption: Adam Smith's spectator theory and the moral constitution of the fair trade consumer', *Journal of International Relations* 13: 263–288.

Watson, M. (2008) 'The materials of consumption', *Journal of Consumer Culture* 8: 5–10.

Wempe, J. (2005) 'Ethical entrepreneurship and fair trade', *Journal of Business Ethics* 60 (3): 211–220.

Wetherell, M. (1998) 'Positioning and interpretative repertoires: conversation analysis and post-structuralism in dialogue', *Discourse and Society* 9: 387–412.

Wetherell, M. and Maybin, J. (1996) 'The distributed self: a social constructionist perspective', in R. Taylor (ed.), *Understanding the Self*. London: Sage, pp. 219–279.

Wetherell, M., Taylor, S. and Yeates, S. (2001a) *Discourse as Data*. London: Sage.

Wetherell, M., Taylor, S. and Yeates, S. (eds) (2001b) *Discourse Theory and Practice*. London: Sage.

Whatmore, S. (2006) 'Materialist returns: practising cultural geographies in and for a more-than-human world', *Cultural Geographies*, 13 (4): 600–610.

Whatmore, S. and Clark, N. (2008) 'Good food: ethical consumption and global change', in N. Clark, D. Massey and P. Sarre (eds), *Material Geographies: A world in the making*. London: Sage, pp. 363–416.

Wiggins, S. and Potter, J. (2003) 'Attitudes and evaluative practices', *British Journal of Social Psychology* 42: 513–531.

Wilk, R. (2001) 'Consuming morality', *Journal of Consumer Culture* 1 (2): 245–260.

Wilkinson, J. (2007) 'Fair trade: dynamic and dilemmas of a market oriented global social movement', *Journal of Consumer Policy* 30: 219–239.

Wilkinson, S. (1998) 'Focus groups in feminist research: power, interaction, and the co-construction of meaning', *Women's Studies International Forum* 21: 111–125.

Wilkinson, S. (2006) 'Analysing interaction in focus groups', in P. Drew, G. Raymond and D. Weinberg (eds), *Talk and Interaction in Social Research Methods*. London: Sage, pp. 72–93.

Winter, M. (2003) 'Embeddedness, the new food economy and defensive localism', *Journal of Rural Studies* 19: 23–32.

Wittgenstein, L. (1953) *Philosophical Investigations*. Oxford: Basil Blackwell.

Yanacopulos, H. (2005) 'The strategies that bind: NGO coalitions and their influence', *Global Networks* 5 (1): 93–110.

Young, I. M. (1990) *Justice and the Politics of Difference*. Princeton: Princeton University Press.

Young, I. M. (2003) 'From guilt to solidarity: sweatshops and political responsibility', *Dissent* 50 (2): 39–44.

Young, I. M. (2004) 'Responsibility and global labor justice', *Journal of Political Philosophy* 12: 365–388.

Young, I. M. (2006) 'Taking the basic structure seriously', *Perspectives on Politics* 4 (1): 91–97.

Young, I. M. (2007) *Global Challenges: War, self-determination and responsibility*. Cambridge: Polity Press.

Young, W. and Utting, K. (2005) 'Fair trade, business and sustainable development', *Sustainable Development* 13: 139–142.

Zavestoski, S. (2002) 'Guest editorial: Anticonsumption attitudes', *Psychology and Marketing* 19: 121–126.

Index

accountability 56, 126
Adorno, T. 16
advanced liberalism 46, 47, 48,
    57, 199
  responsibilization of the
      consumer 40, 41–3
agency, relocating 11–19
American Express 14
Amin, A. 64–6, 68
Amnesty International 157
Anderson, E. 57
anti-consumerism 14, 16
  and ethical consumption, distinction
      between 13
  reflexivity, lack of 94
anti-globalization campaigns 13, 37
anti-obesity agenda 63
anti-sweatshop campaigns
  consumerist turn 15
  justice 6–7, 8, 9
  union-based 14
argumentative interactions, focus group
    research 122
articulation
  of background 77–81
  of consumption and the
      consumer 85–90
  of the ethical consumer 97–107
attitudinal research 117–18
auditing, ethical 98
Austin, J. L. 117, 120
Australia, organic food sector 40

background, articulating 77–81
Bakhtin, Mikhail 120
banking, ethical 13, 14
Barber, Benjamin 30
Baudrillard, Jean 16
Bauman, Zygmunt 16, 30–1, 32,
    39, 61
Beck, Ulrich 31, 32, 34, 36, 39
best-in-sector buying 14
Billig, M. 118
Blair, Tony 98
blame discourses 118
Bloch, Maurice 77–8
Body Shop, the 14
  Ethical Trade Initiative 155
  political consumerism 37
  Roddick, Anita 93
Boltanski, L. 53
boycotts 14, 37, 176
brand names, focus group
    research 129–30, 142, 143
Bristol, Fairtrade City campaign 50–1,
    195–6
  fair trade and 'the politics of place
      beyond place' 191–5
  putting fair trade in place 189–91
  re-imagining Bristol 186–9
  spatialities of fair trade, rethinking
      the 182–5
Bristol Zoo 182, 196
British Wind Energy Association 176
Brown, Gordon 98

Burke, Kenneth 120
business ethics 11, 16
'buycotting' 14
buyers' guides 90–2, 94, 96

Cadbury's 17
Cafédirect 157, 165, 177
CAFOD 163
calculative practices 49–50, 98,
    102–5, 106
charitable donations 19
child labour 141–3
choice
    antinomies of 61–4
    ethical problematization of 'the
        consumer' 27–9, 33
    focus group research 133–4, 148,
        150
    political 33, 36
    problematizing 70–7, 87–8,
        100–1, 114
    problematizing consumption 83–6,
        87–8, 98–102
    sets 101, 102
Christian Aid
    Bristol Fairtrade City campaign 187
    campaigns 176, 177, 193
    choice, consumer 101
    Fairtrade Foundation 163
    influence 18, 106
    political consumerism 37
citizenship
    advanced liberal governmentality 41,
        42–3
    agency, relocating 12, 15
    individualization 29
    moralization of consumption 3, 33
    political consumerism 33–4
    problematizing consumption 84–5,
        98–9, 101
    responsibilization of the
        consumer 41, 42–3, 44
civic engagement and public
    participation 16
    choice as a dimension of 99
    and individualization 30
    problematizing consumption 84, 85
code-oriented morality 55

Cohen, G. A. 5–6, 7
collective activism 85, 96, 166, 177
commodity chains 158
Company Reform Law Bill 167
complementary interactions, focus
    group research 122
complexity of ethical consumption 137,
    139, 141
Conder, Brian 168, 170, 174
consequentialist reasoning 55
constructivist concept of
    personhood 57
Consumer Association 104
contact activism 85, 96, 166, 177
conversation analysis 121, 123
Co-operative Group 88
    Bristol Fairtrade City campaign 187
    differentiating ethical consumers 100
    Ethical Purchasing Index 49
    Fair trade movement 143–4, 156,
        187
    focus group research 143–4
    influence 18, 106
    political consumerism 37
    Shopping with Attitude, news
        coverage 103–4
co-operative movement 14, 37
Corporate Responsibility (CORE)
    Coalition 167
corporate social responsibility
    agency, relocating 11, 14, 16
    Ethical Purchasing Index 99
    responsibilization of the social
        field 42
Coskuner-Balli, G. 136
cost issues in ethical consumption
    127–32, 143–4, 145
Costa Coffee 157
Crewe, L. 119
critical discourse theory 120, 121
Crossley, N. 154, 181
Crowther, Bruce 193
culture-governance thesis 38
culture jamming 14

Davenport, E. 161, 162
Davies, B. 121
Day Sclater, S. 116

Dean, M. 38, 39, 55
  governmentality theory 40, 44–5, 48
de-fetishization of commodities 159
demands of ethical consumption, focus
    groups' reactions to 134
democracy 98
democratic deficit 94
demonstration effects, fair trade
    movement 18
demoralization of consumption 3
Demos 56, 101
Department for International
    Development 185
Department of Trade and Industry,
    Trade Policy Consultative
    Forum 166
desire, consumption as 67
devices to enable ethical
    consumption 19
differentiation of ethical
    consumers 100
discourse
  consumer action 14
  consumer engagement 19
  grammars of responsibility 149–52
    dilemmas of responsibility 137–49
    justifying practices 115–17
    researching the (ir)responsible
      consumer 117–24
    versions of responsibility 124–37
  positioning theory 115–16, 120–2,
    123, 126
  practice-based perspective 77–80,
    116
discursive psychology 120
  methodological approach 55–6
  non-representational tradition 121
  positioning 121
dispersed practices 67–8, 75, 76
distributive justice, global 4–6
downshifting 14
Dreyfus, H. 77, 80
Dulsrud, A. 35

Eagleton, T. 79
Economist, The 12
Ecumenical Council on Corporate
    Social Responsibility 176

Edwards, D. 122
egalitarian theory of justice 5–7
Elliott, A. 29
entrepreneurialism, Fairtrade 178
environmental policy analysis 120
Equop 187
Escobar, A. 190
Esso 14, 37, 176
ethical auditing 98
ethical banking and investment 13, 14
Ethical Consumer 10, 89, 91, 92
Ethical Consumer Research Association
    (ECRA)
  awareness, raising 103
  choice, consumer 101
  ethical consumption activity 14
  Harrison, Rob 10
  mission statement 91
  mobilizing the ethical
    consumer 92–3, 95, 96
  news coverage 105
  problematizing consumption 89
ethical practice, telos of 55
ethical problematization of the
    consumer 20–1, 52–60, 81
  governing consumption 48–52
  grammars of responsibility
    113–14
  political consumerism 33–9
  responsibilization of the
    consumer 39–44
  teleologies of consumerism and
    individualization 27–32
  what type of subject is 'the
    consumer'? 44–8
Ethical Purchasing Index (EPI)
    49–50
  articulating the ethical consumer
    99–100, 102, 103
ethical substance of practices of the
    self 54–5
Ethical Trading Initiative 155, 174
ethical work 55
ethics-oriented morality 55
Eugene, Sherrie 190
Europe, fair trade movement 158
European Fairtrade Association 165
European Union (EU) 188

Fabian Society 56, 88, 101
factual versions concept 126–35, 151
Fairtrade City certification *see under*
    Fairtrade Foundation
Fairtrade Fortnight 105, 188
Fairtrade Foundation (FTF) 156, 157,
    163–4
  choice, consumer 101
  differentiating ethical consumers 100
  Fairtrade Fortnight 105, 188
  Fairtrade Town/City certification 51,
    181–2, 185–6
    Bristol *see* Bristol, Fairtrade City
      campaign
    Garstang 182, 185, 192, 193, 194
  global fair trade network 157
  growth in fair trade markets 157
  influence 18, 106
  mainstreaming fair trade 159
  news coverage 105
  political consumerism 37
  Traidcraft 163, 165
Fairtrade Labelling Organizations
    International 156
fair trade movement 14–15, 17–18,
    179–80
  doing fair trade 170–9
  Fairtrade urbanism 22, 195–7
    'politics of place beyond
      place' 191–5
    putting fair trade in place 189–91
    re-imagining Bristol 185–9
    rethinking the spatialities of fair
      trade 181–5
  focus group research 128, 139, 141,
    143–6, 150
  grammars of responsibility 139, 141,
    143–6, 150
  locating the fair trade
    consumer 153–5
  managing fair trade, mobilizing
    networks 163–9
  re-evaluating fair trade
    consumption 155–63
Fairtrade Town certification *see under*
    Fairtrade Foundation
Fairtrade urbanism *see under* fair trade
    movement

Flyvbjerg, B. 80
focus group research 121–3,
    149–52
  dilemmas of responsibility 137–49
  versions of responsibility 124–37
Food Ethics Council 101
Foucault, Michel
  discourse as a medium of power 120
  governmentality and ethics 27, 39,
    45, 46–7, 51
    ethical problematization of the
      consumer 52–3, 54–5, 58, 59
    responsibilization of the
      consumer 40, 42
  modes of problematization 20, 39,
    199
Foulkes, George 185
Freidberg, S. 98
Friends of the Earth 96, 105, 176
Future Foundation, The 88, 101

Gabriel, Y. 43
Galbraith, J. K. 16
Gap 14, 37, 155
Garfinkel, Harold 52
garment sector 14–15
Garstang, Lancashire 182, 185
  local links 192, 193, 194
genealogical perspective 15, 20, 39
  responsibilization of the
    consumer 40, 41, 43
GeoActivist initiative 168
GeoBars 165
GEPA 156
Giddens, Anthony 31–2, 36, 39, 81
Ginsborg, Paul 71, 72, 73, 74, 76
Global Action Plan 56
global distributive justice 4–6
Good Shopping Guide, The 92
Goodman, M. 85, 159
governmentality theory 39, 44–8,
    58–9, 199
  articulating the ethical
    consumer 98
  ethical problematization of the
    consumer 52–9
  governing consumption and
    governing the consumer 48–52

problematizing consumption 86, 87,
    88–9
responsibilization of the
    consumer 40–3, 59
grammars of responsibility 21–2,
    149–52
  dilemmas of responsibility 137–49
  justifying practices 113–17
  researching the (ir)responsible
    consumer 117–24
  versions of responsibility 124–37
Green Alliance 56, 88, 101
Green Consumer Guide, The 90, 96
Gregson, N. 119
Guardian, The 105

Habermas, J. 33, 52, 53, 152
Hacking, Ian 49, 50, 54, 57
Hajer, M. 28, 86
Hammersley, M. 51–2
Harré, Rom 116, 120, 121
Harrison, Rob 10
Heidegger, Martin 79
Hickman, Leo 105
Hilton, M. 3, 16
historicism 31, 32, 39–40
  political consumerism 34, 35, 36
Hobson, K. 73
Hodges, I. 54, 113
Honneth, A. 53
Horkheimer, Max 16
'How to' guides 90–2, 94, 96

identity 39
  advanced liberalism 41–2
  calculative practices 50
  and collective participation 16
  governmentality 48–9
  mobilizing the ethical consumer 97
  narrative-self 57
  neoliberalism 29
individualistic action 85, 96, 166,
    177
individualization
  political consumerism 34–5, 36, 37
  problematizing consumption 86
  provisioning perspective versus 69
  teleologies of 27–32

individualized collective action 34–5,
    36, 37
influencing, arts of 56–7
information-led understanding of
    consumption 11, 16–17, 56,
    200
  focus group research 131, 140
  governmental shift away from 63
  individualization of responsibility 28
  practice-based perspective 69, 74
infrastructure of consumption 72,
    73
integrative practices 67–8
International Fairtrade
    Association 165
International Monetary Fund 94
interpretative rules and
    regulations 51–2, 123, 134
investment, ethical 13, 14

Jackson, Tim 118
Jacobsen, E. 35
Jacques, M. 28
Jubilee 2000 debt campaigning 167
justice
  egalitarian theory of 5–7
  global distributive 4–6
  responsibility and the politics of
    consumption 4–10
  social justice campaigns 18
justifying practices 113–17

Kant, Immanuel 54
Kraft Foods Corporation 160

labelling 19
Labour Behind the Label 96
Labour Party 98, 100, 166, 176
Lang, T. 43
Lasch, Christopher 30
Lee, R. 10
Lemert, C. 29
Lemke, T. 45
Leyshon, A. 10, 11
liberal paternalism 88
life politics 30–1
Linehan, C. 121
Littler, J. 13, 94

local links, Fairtrade Town/City
　　campaigns 192–4
local networks of global feeling 22,
　　179–80
　doing fair trade 170–9
　locating the fair trade
　　consumer 153–5
　managing fair trade, mobilizing
　　networks 163–9
　re-evaluating fair trade
　　consumption 155–63
Lockie, S. 40
logos, campaigns against
　and ethical consumption, distinction
　　between 13
　mobilizing the ethical consumer 95
　political consumerism 37
Low, W. 161, 162

mainstreaming of fair trade 17,
　　159–60, 161–2, 165, 174
　spatialities of fair trade, rethinking
　　the 184
Make Poverty History campaign 167,
　　175
market campaigns 10, 17, 18
market citizenship 33
market populist paradigm 98–9
Marks & Spencer 157
Marx, Karl 2
Massey, D. 182, 184, 191
McCarthy, J. 121
McCarthy, T. 47, 58
McDonald's 37
media attention 103–5
Micheletti, Michelle 9, 14, 33–4
Miller, Daniel 38, 136, 150
Miller, Peter 49, 86
mobilization
　of the ethical consumer 85, 90–7
　of local networks 163–9
modernity 9, 30–2
modular nature of consumer-based
　　activism 37
moralization of consumption 1–3,
　　16, 33
Morgan, K. 51
Morley, A. 51

multinational corporations 94, 95
Myers, G. 122

narrative-self 57
National Consumer Council 88,
　　101, 102
Neale, B. 81
necessity, consumption as 67
Needham, C. 29
neoliberalism 11, 16, 20, 48, 199
　articulating the ethical
　　consumer 98
　ethical problematization of the
　　consumer 57
　governmentality theory 44, 45, 47,
　　51
　individualization 29, 87
　mobilizing the ethical consumer 94
　responsibilization of the
　　consumer 41, 42
　subject-effects 114
Nestlé 14, 17, 37
New Economics Foundation (NEF) 88
　choice, consumer 101
　Ethical Purchasing Index 49
　shared learning processes 56
New Internationalist 90
New Labour 98, 100
news coverage 103–5
Nicholls, A. 154
Nietzsche, Friedrich 54
Nike 37, 141–2
'No-Logo' campaigns
　and ethical consumption, distinction
　　between 13
　mobilizing the ethical consumer 95
　political consumerism 37
normativity and norms 37–8
　ethical problematization of the
　　consumer 52–4, 57–8
　governmentality theory 47
　interpretation of rules and
　　regulations 51–2
　lay normativity 37, 53, 115
Norris, P. 36

O'Leary, T. 49
O'Neill, Onora 4–5, 124

Office of Government Commerce
     (OGC) 188
One World outlets 156
Opal, C. 154
organic food 40
     Organic Food Week 105
Oxfam 157
     choice, consumer 101
     Fairtrade Foundation 163
     Fairtrade Town/City device 182, 187
     influence 18, 106
     Make Trade Fair campaigns 193
     political consumerism 37
     shops 157, 169

Palmer, Stuart 167
participant observation 78
paternalism
     articulating the ethical
          consumer 102
     liberal 88
     soft 88
Pattie, C. 85, 96, 166
personalization agenda 28, 98
personhood
     constructivist concept 57
     positioning 126
place specificity, fair trade 189–91
pluralism, articulating the ethical
          consumer 98–9
Pogge, Thomas 4, 5, 6, 7
political consumerism 12, 13, 19, 33–9
     activities 14
     fair trade movement 158
politicizing consumption
     agency, relocating 11–19
     justice, responsibility and the politics
          of consumption 4–10
     moralization of consumption 1–3
     practice-based perspective 64–6,
          68, 74
     problematizing consumption 19–23
politics in an ethical register
     beyond the consumer 198–9
     doing responsibility 200–2
politics of shame 93
positioning 115–16
     discursive 115–16, 120–2, 123, 126

methodological approach 55–6
positive buying 14
postmodernity 30–1
post-moralistic approach to politicizing
          consumption 3, 9
poststructuralism 57, 123
power, and shared responsibility 8, 9
practical reasoning 124
     dilemmas of responsibility 137–49
     versions of responsibility 124–37
practising consumption 21, 81–2
     antinomies of consumer choice 61–4
     articulating background 77–81
     problematizing choice 70–7
     theorizing consumption
          practices 64–70
price issues in ethical
          consumption 127–32,
          143–4, 145
Princen, T. 69
privilege, and shared responsibility 8, 9
problematization
     of choice 70–7, 87–8, 100–1, 114
     of consumption 1, 19–23, 107–9
          articulating consumption and the
               consumer 85–90
          articulating the ethical
               consumer 97–107
          consumer choice and citizenly
               acts 83–5
          mobilizing the ethical
               consumer 90–7
     ethical problematization of the
          consumer 20–1, 52–60, 81
          governing consumption 48–52
          grammars of responsibility 113–14
          political consumerism 33–9
          responsibilization of the
               consumer 39–44
          teleologies of consumerism and
               individualization 27–32
          what type of subject is 'the
               consumer'? 44–8
provisioning perspective 69–70
psychology, discursive 120
     methodological approach 55–6
     non-representational tradition 121
     positioning 121

public participation and civic
    engagement 16
    choice as a dimension of 99
    and individualization 30
    problematizing consumption
        84, 85
public policy 27–8, 29
Putnam, Robert 30

quality issues in purchasing
    decisions 129–30

radical consumption 13
Rainforest Alliance 160
Rawls, John 5, 6–7
Reckwitz, A. 67
recycling boxes 19
reflexive modernization 31–2
relationship purchasing 14, 38, 136,
        201
    focus group research 130–2
researching the (ir)responsible
        consumer 117–24
responsibility
    dilemmas of 137–49
    grammars of see grammars of
        responsibility
    responsibilization of the
        consumer 39–44, 59, 200–2
    antinomies of consumer
        choice 62–3
    problematizing consumption 86,
        87, 100
    transnationalization of 93
    versions of 124–37
Ricoeur, Paul 120
risk
    and choice 100–1
    individualization 28, 32
Roddick, Anita 93–4
Ronchi, L. 158
Rose, Nikolas 55, 86–7
    advanced liberal governmentality 39,
        40, 41–2, 43, 46

Sassatelli, R. 88
Sayer, A. 37, 38, 53, 115
scale frames 187, 189

Schatzki, T. R. 67
Schor, J. B. 61
Schudson, M. 3, 30, 33
selfishness, 'age of' 28
Sen, Amartya 101
Seyfang, G. 161
shame, politics of 93
Shamir, R. 42
shared learning processes 56
shared responsibility model 8–9, 10
Shell 138
Shotter, John 116, 120, 124–5, 126
Shove, E. 70, 71–2, 73, 74, 75–6
Shreck, A. 163
simplicity movement 13, 14, 16
slow food movement 13, 16
Smart, C. 81
social connection model of political
        responsibility 7–8
social justice campaigns 18
soft paternalism 88
Soil Association 37, 105
Soper, Kate 10
spatialities of fair trade, rethinking
        the 181–5
Starbucks 157, 159
Stark, D. 53
Stokoe, E. H. 122
Stolle, D. 9
storylines, Bristol Fairtrade City
        campaign 189–90, 192
Strydom, P. 42
subject-formation
    governmentality theory 43, 44–8, 49
    neoliberalism 29
    problematization of consumption 87
subjectivation, mode of 55
Sunstein, C. R. 88
Sustainable Consumption
        Roundtable 102
Sustainable Development
        Commission 102
sweatshops
    campaigns against 6–7, 8, 9, 14, 15
    focus group research 140, 141–3

Taylor, Charles 79–80
Tearfund 164

technologies
 calculative 49–50, 98, 102–5, 106
 of communication 52
 of power 52
 of production 52
 of the self 52
temporal issues in ethical
  consumption 132–4
Tesco 157, 159
textiles sector 14–15
Thaler, R. H. 88
theorizing consumption
  practices 64–70
Thévenot, L. 53
'Third Way' 98–9, 100
Thompson, C. J. 136
Thrift, N. 64–6, 68
Tilly, C. 117
Trade Justice Movement (TJM) 167
Trade Union Congress 158
Traidcraft
 Bristol Fairtrade City campaign 187
 Fairtrade Foundation 163, 165
 GeoActivist initiative 168
 influence 18, 106
 local networks of global feeling
   155, 179
  doing fair trade 170–9
  locating the fair trade
   consumer 154
  managing fair trade, mobilizing
   networks 164–9
  re-evaluating fair trade
   consumption 156, 157
 mobilizing the ethical consumer 96
 political consumerism 37
Transfair USA 156, 158, 159
transnational corporations 94, 95
transnationalization of
  responsibilities 93

Trentmann, F. 43
Triodos Bank 176, 187
Tropical Wholefoods 157
Tully, J. 76

union-based movements 14–15
United States of America
 anti-sweatshop campaigns 6
 ethical trade campaigns 155
 fair trade movement 158
utilitarian conceptualization of ethical
  consumerism 17

value chains 158
Van Vliet, B. 70, 75
Veblen, Thorstein 16, 61
vegetable box schemes 19, 133
Versteeg, W. 28, 86
voluntary simplicity movement 13, 14,
  16
voting metaphor, mobilizing the ethical
  consumer 96
Vygotsky, Lev 120

Warde, A. 67, 68–9, 76
Watson, M. 68
Wessex Water 182
Wetherell, M. 123, 127
Which? magazine 104
Wild Oats Market 156
Wilkinson, J. 154
Williams, C. 10
Wittgenstein, Ludwig 79, 116–17,
  120
women's entrepreneurialism 178
World Bank 94
World Development Movement 163
World Trade Organization 94

Young, Iris Marion 6–9, 10